D0475466

# THE DOOMSDAY MACHINE

# THE DOOMSDAY MACHINE

## THE HIGH PRICE OF NUCLEAR ENERGY, THE WORLD'S MOST DANGEROUS FUEL

MARTIN COHEN AND ANDREW MCKILLOP



palgrave
macmillan

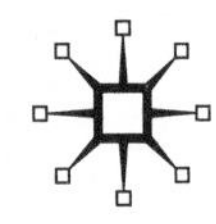

THE DOOMSDAY MACHINE


First published in 2012 by PALGRAVE MACMILLAN® in the United States—a division of St. Martin's Press LLC, 175 Fifth Avenue, New York, NY 10010.

Where this book is distributed in the UK, Europe and the rest of the world, this is by Palgrave Macmillan, a division of Macmillan Publishers Limited, registered in England, company number 785998, of Houndmills, Basingstoke, Hampshire RG21 6XS.

Palgrave Macmillan is the global academic imprint of the above companies and has companies and representatives throughout the world.



ISBN 978-0-230-33834-0

Library of Congress Cataloging-in-Publication Data

Cohen, Martin.
The doomsday machine : the high price of nuclear energy, the world's most dangerous fuel / Martin Cohen, Andrew McKillop.
p. cm.
Includes bibliographical references and index.
ISBN 978-0-230-33834-0 (hardback)
1. Nuclear industry. 2. Nuclear energy—Economic aspects. 3. Nuclear energy—Environmental aspects. 4. Nuclear power plants. 5. Nuclear accidents. I. McKillop, Andrew. II. Title.
HD9698.A2C567 2012
333.792'4—dc23

2011037618

A catalogue record of the book is available from the British Library.

Design by Letra Libre, Inc.

First edition: March 2012

10 9 8 7 6 5 4 3 2 1

Printed in the United States of America.

# CONTENTS

# FOREWORD

*Steve Thomas*

**TO LISTEN TO MANY ENERGY EXPERTS, LET ALONE THE** nuclear industry lobbyists themselves, nuclear power is both an essential and inevitable part of the energy mix. Doing without nuclear is said to be a romantic dream. Yet, on the contrary, it is the pursuit of nuclear energy that seems to be the stuff of idealist dreams. Instead, one of the most puzzling questions about nuclear power is why it is still being pursued. Like a lot of new technologies, the early versions were not economically feasible, but there was an expectation that, as with any successful technology, over time, real costs would fall substantially and technical performance would improve markedly. These benefits would be brought about by a combination of: "learning"—the more you do something, the better you get at it; scale economies—the larger the output of each facility, the lower the cost per unit of output; economies of number—the more units you produce, the cheaper each unit is; and general technical progress. Most people's expectations are based on experience with consumer technologies like mobile phones and televisions, where today's versions are barely recognizable compared to those of ten to twenty years ago. This sort of development has also been the rule with power generation technology for more than a century. New technologies are introduced and improve rapidly until, when the technology is fully mature, performance begins to plateau and, perhaps, new technologies come in to replace the old one.

Nuclear power has never followed this expected path. Real costs have consistently and substantially risen over the 60 years that the technology has been pursued. Technical performance in terms of the number of kilowatt-hours (kWh) of power a given amount of uranium produces is little better in the newest designs than it was in the 1960s. Yet the technology is still being pursued with apparently as much fervor as ever. Part of the explanation for this paradox lies in the myths that Martin Cohen and Andrew McKillop expose here—myths that have grown up around nuclear power and that have proved highly durable despite the formidable body of evidence that they are just that, myths. Two of these myths are personified in those well-known phrases from the 1950s: "Electricity too cheap to meter" and "atoms for peace."

This strange history begs the questions: How can we explain the apparently implausible survival of these myths? How has it been possible to maintain the prominence of the nuclear option despite the evidence of the failure of the technology? Who has been promoting nuclear power?

The main ways for selling the nuclear pill have been claims that the world faced serious problems that only an urgent and rapid deployment of nuclear power could solve and that the reason nuclear power had not fulfilled the promises made for it was that, earlier, it was being done wrong and that this time it will be done better and all will be well.

Another explanation for nuclear power's otherwise inexplicable survival as an energy strategy is that rival technologies continue to fall in and out of fashion. At every twist and turn, nuclear is there, offering to solve all the world's urgent problems. In the 1960s, when nuclear power first started to be strongly promoted globally, the rationale was that reserves of fossil fuels, such as coal, were declining steeply and prices would rise, making them prohibitively expensive as power generation fuels; thus, nuclear power would be economical. The first oil crisis of 1973–1974 gave an added twist to this concern with the public recognition that oil reserves were highly concentrated in the Middle East countries, thus making the

West apparently vulnerable to political blackmail. This message was reinforced again by the Iranian revolution in 1979 when oil supplies again seemed at risk. By the early 1980s, public environmental consciousness was much higher with the effects of acid rain, supposedly resulting from the burning of fossil fuels, which became a major concern. Switching to nuclear was portrayed then too as a way out of the problem. Yet the slump in fossil fuel prices of the mid-1980s and the way in which the acid rain problem was dealt with by relatively simple technical fixes, combined with the impact of the 1986 Chernobyl disaster, left nuclear power deep in the doldrums. However, new ammunition for the nuclear industry was given by the first Iraq War, the gradual transition of global warming from environmentalists' fantasy to the stuff of scientific "consensus," and the growing credence given to the "peak oil" theorists. According to this theory, oil production was close to its peak and must soon begin to decline steeply, bringing severe economic disruption if oil use was not drastically reduced. It was this fertile ground and these arguments that fed the nuclear renaissance that began to be promoted by the nuclear industry, ever anxious to seize any chance to gain political support, around the turn of the millennium.

Another recurring feature of attempts to restart orders for new nuclear has been a claim of great urgency for these commands. The nuclear industry's last-resort argument has always been that if the nuclear program was not relaunched immediately, the lights would go out. As a result, it was increasingly common to hear that normal democratic planning processes had to be accelerated or even overridden and safety assessments had to be done with all possible speed. So far these threats have all seemed to be crying wolf. If we look at the realistic time scales for nuclear investment—at least ten years from decision in principle to invest to first power and with ample scope for delays—whenever urgent action is really needed, nuclear is the last option to look at.

Yet, for many people, if global warming and fossil fuel depletion really are the main threats then, intuitively, nuclear still appears on the face of it to be the only energy source that could make a large contribution to solving the problem. However, some simple arithmetic easily disproves this. In the United Kingdom, electricity accounts for only about 20 percent of our energy needs. We get about 15 percent of our electricity from nuclear, so, at present, a mere 3 percent or so of the energy used in the United Kingdom is nuclear. If we were to increase the share of nuclear in electricity to the near 80 percent level of France, we would still be getting only one-sixth, or 16 percent, of our *energy* from nuclear, with almost all of the rest still having to come from fossil fuels. So, to be effective, electric vehicles and electric space and water heating would have to be deployed on a massive scale to get the nuclear share up to, say, 50 percent. Leaving aside the economics and how people living in the chilly British Isles would be able to afford to heat their homes with electricity, the number of nuclear stations in the United Kingdom would have to increase from just ten now to close to 200. The problem of where to site such a large number of nuclear stations would not be the only one. At that level of deployment worldwide, natural uranium supplies would run out very fast while short-term attempts to use fast reactors instead would likely require large quantities of plutonium to be shipped around the world, creating a significant proliferation and terrorism risk. And then there would be the disposal of the large quantities of waste produced . . . the list of issues goes on.

## WE HAVE BEEN DOING IT WRONG BUT NEXT TIME IT WILL BE OK

The nuclear industry has been remarkably reluctant to admit to any failings on its part. When things go wrong, it is always the fault of someone else: Luddite environmentalists; obstructive regulators; an ignorant public; weak-willed politicians motivated by narrow political motives; or just plain bad luck, as with the Fukushima disaster. Who would have guessed it? An earthquake and a tsunami all at once! (Although plenty of people did "guess" it, and indeed specific warnings were given about Japan's nuclear industry risks.) As a result, it is argued, we have basically been doing it wrong. Designs should have been standardized, series ordering undertaken, the scope for public opposition strictly limited, and expertise concentrated in a few highly competent companies. However, in fact, this is just what one country, France, has already done. Thus, we can test this theory by looking at the experience there.

In 1975, the French government launched a massive program of nuclear orders. It concentrated its nuclear expertise into one utility, one reactor vendor, and one fuel cycle company, all with strong and mostly total state ownership and coordinated by a single ministry, the Ministry of Industry. The scope for opposition was minimal, and the safety regulator was fully supportive. The result was that between 1975 and 1990, France ordered 58 nuclear power plants in batches of up to ten plants of effectively identical design, allowing efficient production-line manufacture of components, maximization of any learning process, and profiting from the general technical advances of that period, particularly in information technology. Surely under these ideal conditions, costs would fall dramatically and efficiency would improve markedly.

And indeed, so convinced are most people that France sets a fine example for all other countries to follow that they do not bother to look at the evidence. Far from achieving dramatic reductions in costs over the period, recent estimates suggest the real construction cost of the last plant was up to *three times* that of the first. Add to which, in a recent government-commissioned report on the state of the French nuclear industry, François Roussely, a former chief executive of the French utility Électricité de France—usually abbreviated to EdF—stated that whereas all around Europe, the reliability of nuclear power plants was improving, in France, it was getting worse. In fact, France had always lagged behind countries such as Germany, Switzerland, and the Netherlands by the reliability criterion. However, in the context of the nuclear renaissance, it is the evidence from France's own Flamanville plant that is most damaging to the country's reputation. That said, in fact Flamanville was not the French vendor Areva's first attempt at building a new nuclear reactor. In 2004, Finland had bought the first of the newest-generation design, the European pressurized reactor (EPR, also dubbed, essentially for marketing purposes outside Europe, the evolutionary power reactor), from Areva. A lot had hung on this order for Areva, and there was an expectation that this would be a showpiece project that would go like clockwork to shop-window the new design. In the event, these hopes were quickly confounded. The Olkiluoto plant was expected to take four years to build and cost €3 billion. Yet, in 2009, after four years of construction, it was at least four years from completion and the cost estimate had all but doubled.

So the script was quickly revised and the problems both excused and dismissed as inevitable in the first of a kind, especially in a country whose most recent nuclear construction experience was 25 years earlier. Instead, in best nuclear industry tradition, the story was that next time, in the safe hands of the world's most experienced nuclear utility, EdF, and on home soil with access to all France's skills, the second order, Flamanville, *would* go smoothly. As to that, work started in December 2007 with an expected five-year construction program costing €3.3 billion. By July 2011, it was clear that things were actually going worse than at Olkiluoto. The plant was four years late, estimated costs had doubled, and, after two fatal site accidents, EdF was having to rethink its entire construction process. Clearly, the idea that there is a "right" way to do nuclear as exemplified by France is another myth.

In its humbler moments, the nuclear industry does acknowledge that it has made errors, yet the rider is always that the lessons have been learned, new designs will overcome previous problems, and the next time, everything will go smoothly. At the same time, the running sores of the industry—the failure to make meaningful progress with waste disposal and decommissioning of retired plants—are portrayed as purely problems of NIMBYism and the lack of any compelling reason to actually carry out these processes. The waste disposal and decommissioning techniques are said to be well known and well established on a small scale. However, 35 years ago, in a report for the UK Royal Commission on Environmental Pollution, Lord (then Sir Brian) Flowers wrote: "There should be no commitment to a large programme of nuclear fission power until it has been demonstrated beyond reasonable doubt that a method exists to ensure the safe containment of long-lived, highly radioactive waste for the indefinite future." This report, which has stood the test of time remarkably well, was ignored.

## WHO IS THE NUCLEAR LOBBY?

Perhaps an even more difficult question to answer is who the nuclear lobby is. Thirty to forty years ago, it was easy to construct a picture of the nuclear industry being pursued on grounds of self-interest. All major countries had a national nuclear research and development (R&D) agency with high prestige and with the ear of government. Nuclear reactors were sold by highly influential "national champion" companies: General Electric (GE) and Westinghouse in the United States, Siemens in Germany, Ansaldo in Italy, and so on. Nuclear was a technology that only the largest and most sophisticated companies could take on, and pursuing nuclear power only seemed to cement their position. The electric utilities took no commercial risk building nuclear power plants. However badly things went wrong, they simply passed the bills on to their captive consumers. Pursuing nuclear power added to their prestige and gave them much more interesting and challenging tasks than building boring old coal plants.

Thirty to forty years later, all these conditions have gone. The nuclear R&D organizations have either been closed down or are shadows of their former selves. The large companies have disappeared (GEC), exited nuclear power (Siemens), sold off their nuclear division (Westinghouse), or drastically scaled back their nuclear activities (GE and Ansaldo). Electric utilities generally no longer have captive consumers and cannot expect to pass on all the costs they incur to their consumers, so they can no longer afford to indulge in "luxury" technologies. Only in France do these conditions still exist. There, EdF remains a de facto monopoly; Areva, the reactor vendor, is a major company; and the R&D organization, the Commissariat à l'Energie Atomique, retains a lot of its prestige.

Today, despite the apparent withering of the forces for nuclear, the lobby for nuclear is as strong as ever. However, its champions are now at the top of government. In the United States, both Republican presidents, such as George W. Bush, and Democratic ones, such as Barack Obama, have shown themselves determined to press on with their nuclear aspirations no matter how badly things go. Similarly in the United Kingdom, Tony Blair, then Gordon Brown, and now David Cameron have taken a very personal interest in the nuclear program. But their Canute-like stance, in attempting to order the tide to turn, cannot prevail. Political backing has its advantages, but these advantages can be overstated. In the 1980s, Margaret Thatcher and Ronald Reagan were known as champions of nuclear power, but they did little or nothing to try to halt its decline. Thatcher allowed her program of ten new reactors, announced in 1979, to be cut back to just one order. Reagan presided over a massive wave of cancellations of old orders (about 80), and no new orders were placed under his watch. The strong support of Silvio Berlusconi and Angela Merkel has not been enough to prevent Italy and Germany from turning determinedly away from nuclear. It is more the economic realities of rapidly escalating costs and insurmountable financing problems than the random acts of Mother Nature that will mean that the much-hyped nuclear renaissance will one day be remembered as just another "nuclear myth."

*Steve Thomas, Professor of Energy Studies at Greenwich University in the United Kingdom, is a researcher in international energy policy of over 30 years' standing and is the author of numerous influential reports and papers, including* The World Nuclear Industry Status Report 2009.

# INTRODUCTION AND OVERVIEW

*From its jungle of steel pipes and stairwells to its gracefully tapered cooling towers, Calder Hall atomic power plant looms like a phantom of the future over a bleak northern landscape strewn with relics of the past. Inside, white-smocked technicians check their controls in vast, immaculate steel and concrete chambers. The silence is broken only by the low hum of the blowers which force carbon dioxide gas through the hot reactor and on into the heat exchangers, where water is turned into superheated steam for driving the turbines.*

—*Life* magazine, 1956

*The Queen of England opening the new atomic pile at Calder Hall, the world's first full-scale nuclear power station. (Press Agency)*

**IT IS A GRAY OCTOBER DAY IN 1956, IN A DRAB NORTH**ern town in England, and Her Gracious Majesty, Queen Elizabeth II, is opening the world's first civilian nuclear reactor. She stands on an enormous, specially constructed pagoda, surrounded by flunkies. Over her head, attached precariously on a large steel pipe, is quite possibly the largest royal coat of arms ever made, bearing the familiar British motif of a lion eating a unicorn.

Two technical marvels—a fine new coat of arms and a nuclear reactor—combining at one pivotal moment in history. No wonder the Queen felt confident enough that day to declare that Britain was now aiming "to produce enough atomic plants by 1975 to satisfy two-thirds of the country's needs."

Not that it had any realistic chance of happening. And curiously, the official program for the day put a rather different, and surprisingly frank, slant on the real purpose of civilian nuclear power, saying: "Calder Hall was built as a requirement for more military plutonium and as an experiment to investigate the possibilities of adapting nuclear energy to the production of electrical power quickly, cheaply and safely."

In other words, for Britain, nuclear power was developed in order to produce material for nuclear weapons; electricity was merely a useful by-product. The concept of nuclear electricity meeting civilian needs was only so much hogwash then. Yet the same promises are being made again today, as nuclear power spreads around the world with great rapidity yet with countries rarely purchasing more than a handful of reactors.

It is now half a century after Calder Hall, and we are still grappling with the realities of nuclear power: nuclear waste and nuclear proliferation.

## SEDUCTION AND BETRAYAL

In recent years—at least up until March 11, 2011, the day of the Japanese tsunami and the Fukushima disaster—nuclear power has experienced a glorious renaissance. It was again being touted as producing huge amounts of "clean, safe, and cheap" power as well as being uniquely sustainable; its proponents claim there is no basic shortage of uranium fuel. In short, it was once again put forward as the energy solution to suit almost every country, anywhere. This line of promotion, now moving fast from the old-nuclear North to the lower-income global South, is so strong that it is easy to forget the bottom-line costs of nuclear power, which remain unchanged. Even now, nuclear power supplies only about 15 percent of the world's electricity and barely 6 percent of its commercial energy needs. Six percent! And for that, the world must hold its breath as earthquakes threaten to topple Japanese reactors like so many misaligned dominos.

The reality is that the supply of atomic electricity is still dwarfed by the supply provided by coal and natural gas, equaled by world hydroelectric production, and these days, at least in theory, rivaled by *potential* wind-power capacity. As Myth 1 explains, today, as for many years past, coal supplies nearly half (45 percent) of the United States' electricity, and overall just a shade under 30 percent of the world's primary energy. Despite this, in almost every country—usually for reasons completely unrelated to its ability to deliver electricity—there is almost universal political support for nuclear power. This support spans almost all stages of economic development and national income. It is seen as a friend to rich and poor alike, happy to partner with governments of all colors, whatever their politics or form.

Should we attribute this open door to national prestige, the appeal of high tech—or the ever-spreading influence of privatized nuclear business? No matter the reasons, the appeal of the atom has worked its way deep into the public consciousness, democratizing the illusions that have always made it irresistible to political and corporate elites.

In this new nuclear renaissance, atomic technology is no longer held at arm's length because of its link with nuclear weapons; rather it is enthusiastically embraced for its associations with science, the Big Bang, quantum physics, and the hunt for ever more elusive fundamental particles. The atom has always been a rich vein for the mythology of popular science, enhancing its mystique for decision-making elites, well-read citizens, and academics alike. And so the nuclear sales campaign mixes popular science images with nuclear myth and reality, blending image and illusion, weaving them together like nuclei fused in the Sun.

## THE HAZARDOUS ATOM

The one thing everyone, even politicians, knows about atomic power is that it is hazardous. But how hazardous? Almost no one can fully face the question. They can see the gleaming nuclear iceberg but only surmise the deadly risks below.

Nuclear weapons are openly terrifying. Most people rightly continue to associate the atom bomb with the Armageddon-like images of Hiroshima and Nagasaki, despite this cathartic display of the atom's might now already having receded more than two generations. Even so, the cold fact is that nearly twice as many people were killed in the January 2010 Haiti earthquake and five times more people killed by the 1994 Rwanda genocide, attacked by mobs armed with machetes and rocks, than in those two cities combined. Yet these events seem somehow less striking, in the same way that a yearly toll of car accidents is less appalling than a lone gunman on a shooting spree. Part of their power is that nuclear weapons still terrify.

Nuclear *power,* however, is openly benign—at least until the recent events in Japan. Before the meltdown, it seems as if we heard only about the positive side, the "virtuous application" of atomic knowledge. Even if, unfortunately, each and every day the world's civil power reactors increase the volume and quantity of the deadly, long-lived radioactive materials produced inside their cores. Their most dangerous excrement must be extracted, separated, and then stored away from all contact with all living things for many times the life span of any civilization that has ever existed on this planet. But this toxic legacy should be unsurprising when one recalls that these power plants are also the Doomsday Machines essential to producing the deadly materials for weapons of mass destruction.

Disposing of the waste is simply impossible, although this fact is often denied. Instead we are told about the few "permanent repositories" that exist, and of the others that are planned, but that progress is slow because they cost so much, the residents complain, and the geology has to be studied very carefully. We are not, however, told the real reason they are not being built: because they will not make any difference in the long term or even in the medium term; or that vast mountains of low-level waste are already out of control and creating telltale rises in rare cancer and leukemia rates. (This is the subject of Myth 7, where the health debates are described in greater detail.) This denial leads to ever more generous extensions to reactor lives in the hope of putting off the costs of phasing out and decommissioning nuclear plants once their useful lives are over.

All of which reveals another basic facet of nuclear power that separates it from all other energy sources and systems: Its risks are open-ended not only in cost but also in time. For this reason, nuclear threats and hazards are always minimized or denied, preferably the latter.

Take one aspect of the nuclear mythos, repeated so often that it is widely believed. Nuclear weapons as we currently think of them are not, in fact, so very high tech. The world's first and only nuclear bombs ever dropped were produced using the industrial technology of the early 1940s, by our grandparents and great-grandparents. The Manhattan Project's bombs, to be sure, were built by colleagues and rivals of Albert Einstein, which may carry the gloss of cutting-edge physics to the uninitiated, but the technology of the time was primitive—certainly when compared with the materials, techniques, and machines available to today's nuclear proliferators.

The Manhattan Project technology was light-years away from what we consider high tech today. If the Apollo rockets struggled to the Moon using computers that would fail to run a washing machine today, the first nuclear bombs had to be constructed in a pre-computer world. Instead, the project drew heavily on US automotive technology of the time—the Fordism of the first mass-produced cars. Materials and techniques included cat gut and piano wire, cast iron and plywood, animal glues, vegetable-based paints, handmade die cast molds, and hand-beaten tin sheets.

### THE DOOMSDAY THREAT OF THE NUCLEAR RENAISSANCE

Political deciders, corporate elites, and mainstream media have swung to almost total support for nuclear power. In the decade between 2010 and 2020, as many as 200 to 250 reactors, each equivalent in size to the industry-standard model of 900 megawatts, may be built. Most will be built in emerging and developing countries, in what the industry calls "the nuclear renaissance."

Industry promoters say reactor construction could rival its previous high-water mark, during the mid-1970s and early 1980s, when the industry averaged one new reactor coming online every 17 days for 10 years. Each new reactor will produce as much as 50 kilograms of plutonium and over 30 tons* of other high-level radioactive wastes each year inside its fuel rods and reactor assemblies. These materials will need to be extracted, reprocessed into reactor fuel, or stored in "plutonium repositories" for the plutonium, and "high-level waste repositories" for the other materials.

What this means is as simple as it is devastating: Each large industry-standard reactor contains a radiological inventory equivalent to 150 or 250 times the total radiation release of the 1945 Hiroshima atom bomb.

---

* A ton, also known as a long ton, is 1,000 kilograms, or 2,200 pounds.

No one understands radiation. Certainly Polish-born physicist Madame Curie did not when she first laboriously collected radioactive materials in a shed in Bohemia. After all, human beings live in a world of objects, and radiation is invisible; it can travel through walls and generally defies all the normal rules of physics.

Indeed, as far as science goes, invisible radiation has been around only since about 1886, when German physicist Heinrich Hertz demonstrated experimentally the existence of radio waves. (The radio frequencies are named after him.) Hertz played a key part in the development of not only radio but TV and the investigation of radiation in general, such as X rays.

Hertz was particularly intrigued by the fact that radio waves could be transmitted through different types of materials, including steel, and yet sometimes were reflected by others (an observation that would lead, much later, to the development of radar). However, Hertz did not realize the practical importance of his experiments. He is said to have commented, "It's of no use whatsoever [ . . .] this is just an experiment that proves Maestro Maxwell was right—we just have these mysterious electromagnetic waves that we cannot see with the naked eye. But they are there."

James Clerk Maxwell being, of course, the Scottish originator of the Laws of Electromagneticism. Hertz died in 1894, at just 37 years of age. A year later, in September 1895, Italian inventor Guglielmo Marconi sent the first radio signal over a distance of just under a mile. In 1901, he sent a signal across the Atlantic.

And on November 8, 1895, German Wilhelm Conrad Röntgen at the University of Würzburg discovered a new kind of radiation, which he called X rays. Once again, the ability of the radiation to pass through opaque material that was impenetrable to ordinary light created a great sensation. Röntgen himself wrote to a friend that he worried that people would say, "Röntgen is out of his mind." However, on January 1, 1896, he made his first announcement of his discovery, "and now the Devil was let loose."

Indeed it was, but perhaps not in the sense he meant. For if, in medicine, researchers like Marie Curie would use it "for good," in physics, the discovery of the mysterious "uranium rays" led inexorably to the bomb. In fact, it would take just 50 years to get from those first, curious inquiries to the man-made sun in the sky over Hiroshima, melting buildings, trains, and people below. Fifty years after that, nine nations have atomic weapons. As for the next few years, at least 45 nations have the technological and industrial capacity to build nuclear bombs if they want to, with another 15, it seems, on the verge. Proliferation is yesterday's story. Atom bombs are already everywhere.

### AMAZING FACT: THE NUCLEAR NINE

**5** declared nuclear weapon states: all permanent members of the United Nations Security Council

**3** de facto nuclear states: India, Pakistan, and North Korea

**1** "open-secret" nuclear arsenal: Israel

This high number of "new nuclear" states underlines the fact that the limits on the spread of the bomb are not technological but political. Realpolitik, not some murky ability to access quaint 1950s-style atom secrets, determines whether a country will own or acquire the bomb. The fact that so many nations *can* build the bomb, but have not, is itself an important part of the story—"the dog that did not bark," as Sherlock Holmes might say.

Nonetheless, the number of countries that can or might build nuclear weapons continues to grow, and that is not even considering the headline-grabbing potential of "terror organizations" or organized crime syndicates (whose weapon of choice would very likely be downsized "dirty bombs"—regular bombs with a dash of radioactivity, enabling them to kill a thousand times more people).

When Pierre and Marie Curie discovered radioactivity in those very early days, they did so by laboriously extracting radium from a vast slag heap of the mineral pitchblende, the most common uranium mineral, left over by a mine in Bohemia. They worked in a large shed with a glass roof that let the rain in, making it muggy in the summer and drafty and cold in the winter. Marie carried out the chemical separations, an early version of the process now known as uranium enrichment, while Pierre undertook the measurements after each successive step. Physically, it was heavy work. Marie processed about 40 pounds of raw material at a time. First she had to clear away any pine needles and visible debris; then she had to cook the whole ghastly stew. "Sometimes I had to spend a whole day stirring a boiling mass with a heavy iron rod nearly as big as myself. I would be broken with fatigue at day's end," she wrote later.

The only furniture in the shed was some old pine tables where Marie worked with her costly radium fractions. Since they did not have any cupboards in which to store their precious products, they arranged them on tables and planks. Marie recalls the joy the couple felt when they came into the shed at night, seeing "from all sides the feebly luminous silhouettes" of the products of their work. (Marie liked to keep a little radium salt by her bed that shone in the darkness.) After thousands of crystallizations, they finally—from several tons of the original material—isolated a tiny, glowing portion of almost pure radium chloride and were thus able to determine radium's atomic weight—it is 225.

Not long after, Marie and Pierre were invited to the Royal Institution in London, where Pierre gave a lecture. Before a packed and astonished auditorium, he showed how radium could affect photographic plates firmly sealed against light in paper; how the substance gave off a mysterious heat; and, most impressive of all, how it glowed in the dark. He described medical tests he had tried out on himself, including wrapping a sample of radium to his arm for ten hours and then studying the wound it caused, which resembled a burn but lasted much longer; indeed, it left a permanent gray scar. In connection with this experiment, Pierre mentioned the possibility of using radium in the treatment of cancer. But members of the audience noted that Pierre's scarred hands shook as he talked, so much so that at one point he spilled a bit of the precious radium preparation.

In actual fact, Pierre was very ill. His legs shook so much that at times he found it hard to stand upright. He was in a great deal of pain. The only treatment doctors could prescribe for such symptoms was strychnine. Marie, too, had strange and disquieting symptoms. The skin on her fingers was cracked and scarred. Like Pierre, she suffered from constant fatigue. The two discoverers of radioactivity had no idea that radiation could have a detrimental effect on their health, any more than the children playing near Calder Hall's new power plant would have a century later. How can something invisible harm you? Out of such natural but fatal intuitions, the nuclear industry has survived and flourished.

If today, at the Bibliothèque Nationale in Paris, you want to consult the three black notebooks in which the Curies recorded their work in 1897 and after, you have to sign a certificate that you do so at your own risk because the notebooks are still radioactive. They will be for a long time too. In fact, it will take 1,620 years before the radioactivity of the notebooks will decrease by a half.

Pierre died in 1906, but it was not from radiation poisoning—he was run over by a horse. Marie Curie survived much longer, but on July 4, 1934, she succumbed to the slow poison of the atom and died of leukemia.

A millennium and many reactors later, nuclear hazards are still little understood. Problems include the relentless increase in the number of reactors and therefore of the amounts of deadly radioactive materials that exist and must be stored or transported around the globe. There is also the small matter of depleted uranium weapons and dirty bombs of various sizes—from a beer-can-sized canister of plutonium oxide in a city center trash bin to a mishap at a full-blown nuclear reactor. And there are endless doomsday scenarios possible from just one worst-case core meltdown, including the rupture of the main containment features and release of high-level radioactive material.

Speaking of nuclear mishaps, Chernobyl, like Fukushima in 2011 and many other nuclear power near misses, is usually explained away as an accident ascribable purely to human error with perhaps a dash of institutional negligence. But in the case of Chernobyl, the core meltdown had a special significance—and an especially fearsome radiation cloud—because the reactor was very large. Even as suicide helicopter pilots valiantly managed to put out the fire in the reactor core and then entomb the broken reactor in a coffin of boron, sand, and concrete, a cloud of invisible radiation quickly covered most of western Europe, stretching from the beaches of southern Spain to the highlands of Scotland. No one knows how many people died as a result. The International Atomic Energy Agency (IAEA) estimates it at 9,000—and reminds us that people die from coal-mining too. Greenpeace thinks the number is more likely at least ten times as many—maybe 100,000. But whatever the number, when we dig a little deeper, we find an even more frightening fact: This disaster *will go on killing* for decades, slowly but surely. Only time and the countdown to zero of radionuclide half-lives will conclude this process, as radiation hot spots and the constant recycling of highly radioactive materials approach the natural background radiation level.

The terrible concomitant effect of nuclear power—*the long-life radiation hazard*—is one problem that remains unresolved.

## A RENEWED CALL FOR ATOMS FOR PEACE

The spread of the bomb is the inevitable price we pay for the spread of nuclear power. But there are other veils to pierce around nuclear power—beatific ones that promise a cornucopia of cheap energy and earth-bending power. For example, in Dwight D. Eisenhower's "Atoms for Peace" speech at the United Nations (UN) on December 8, 1953, he cited nuclear power's potential to build canals hundreds of miles long and irrigate deserts—a green dream still echoed today by prominent environmentalists.

At least until the Fukushima setback, the nuclear lobby and its friends in government triumphantly announced a constant flow of new and massive orders for new power plants and were especially pleased to sell to the new, emerging economies of the South, almost every few months. Outside China, India, and the Gulf petrostates, three of which have already announced spectacular and massive nuclear power plant orders, nearly all the new-nuclear countries are still nonindustrial and poor, or very poor. No matter that these are lands periodically wracked by food shortages, civil strife, and unrest, or that they have recently experienced civil war and border conflicts with neighboring countries. No matter that most of them have tiny domestic demand for electricity (sometimes 50 times less than the average electricity consumption per capita of the countries of the Organization for Economic Cooperation and Development). Forget that their power transmission and distribution infrastructures are often frail and unreliable. Or that, of course, they have no previous experience with nuclear power. Because the nuclear image and the reality on the ground are two vastly different worlds. The reality is that the national infrastructure spending needed to upgrade to the atom and integrate huge new power plants is itself beyond these countries, but pretending otherwise offers juicy pickings for companies and corporations in the rich world. The reality is that the nuclear stampede of the new-nuclear South will be straight into a debt trap.

How could low-income countries such as Sudan, Nigeria, Ghana, Egypt, Indonesia, the Philippines, and others afford their nuclear medicine? Only all kinds of creative financing methods are offered to fill the void. (This aspect we will discuss further in Myth 5, "Nuclear Geopolitics Trumps Geopolitics.")

But this very practical problem is hardly admitted and scarcely commented on. Indeed, only the stories about bombs make headlines. Perhaps that suits the nuclear industry, because in that arena, at least, "safeguards" can be proposed. The debt and inequality that nuclear power brings in its wake are much harder to whitewash. Sales teams from the world's nuclear industry, which is highly restricted with fewer than ten major supplier nations, constantly scout the globe for new and bigger contracts. They have already struck gold in India, home to 300 million people below the UN poverty line of $2 a day—but also the world's most massive nuclear market. Incredibly, in a land unable to afford water and sanitation, let alone basic education and health care, sales of reactors and equipment (excluding fuel supplies!) could exceed $175 billion over the next 20 years.

## MILITARY AND CIVIL PROLIFERATION

From the mid-1960s until the early 1980s, first India, then Israel, and finally South Africa and Pakistan "acquired" the nuclear materials needed to make a bomb. If we add in the activities of at least two or three other countries, such as Argentina and Brazil, we can estimate that as many as 1,000 nuclear warheads were built under the not-so-steely gaze of the

Nuclear Non-Proliferation Treaty (NPT). All these countries used civil-source nuclear technology to develop their atomic weapons. In some cases, the programs took as long as 12 to 15 years; in other cases, they took considerably less. These sorry examples of "clandestine" nuclear weapons development show that the NPT had little impact on events. The treaty powers, in fact, followed developments at a respectable distance, measured in years or even decades—as witnessed when India was welcomed back into the fold in 2008, after being subject for a suitably short period to the Great Powers' disapproval.

In fact, because politics is paramount, many of these "illegal" bombs have been dismantled, but considerable numbers of nuclear devices in this clandestine category, which also includes atomic mines, depth charges, and other ingenious explosive devices (with an explosive potential equal to the Hiroshima weapon), still exist. As of 2010, there were probably 800 such devices produced and held by nations outside or in defiance of the treaty.

So despite the NPT and the UN watchdogs, notably the IAEA, and the increasing sophistication of surveillance technology of all kinds, "turning the screwdriver" (i.e., very quickly switching reactors from being civilian to being military) is eminently feasible.

In the coming decade, as many as 20 to 25 nations could take the same proven civil-to-military route to gain nuclear weapons. By 2020, the rogue and clandestine nuclear powers—that is, weapon states outside the NPT—might have 1,500 or more Hiroshima-size atomic weapons to hand. Countries, or rather political leaders, are inexorably sucked toward nuclear weapon status one by one, as their neighbors achieve it. And "status" is the key word, after all.

This is true because the five, and only five, permanent members of the UN Security Council all have the weapons. During the Cold War, they all said they needed their bombs for protection. But now they are all friends. Yet the weapons seem not to have diminished in importance. Politicians in both France and Britain have trotted out their Security Council seat as one good reason to keep their bombs. Evidently, like a highwayman deprived of his gun, they fear that without being nuclear, they might lose that very status and enhanced political clout.

## A QUICK FIX?

Instead, the rapidly industrializing South's massive push to developing atomic energy is the fundamental pillar of the industry's vaunted "nuclear renaissance." It is a fantasy called the nuclear quick fix, where the friendly atom almost instantly solves every problem (constant power brownouts and blackouts) for developing and emerging economies engaged in a flat-out race for economic growth. By 2020, on current plans, this renaissance could add another 250 giant-size reactors to the current world fleet of 439 civil reactors and about 200 research and military reactors. As many as 50 to 60 of these reactors will be built in poor or very poor countries within the emerging economy group, massively raising the volume of nuclear waste—and potential bomb-making materials—circulating outside the relatively well-supervised OECD countries and worldwide.

The truth is that new reactors are expensive, usually costing several billion dollars each. If built, in reality, they will do little or nothing to solve power shortages in the South or reduce total carbon dioxide emissions, let alone limit supposed global warming, yet they certainly will entail real dangers, risks, and hazards—economic as well as humanitarian. The economic challenge, like the security challenge, remains unanswered. As does another problem: Where will the fuel for all the new reactors come from?

Uranium mining has rapidly shifted from North to South in the last decade, with large-scale and long-lived pollution, environment damage, and human health damage in countries including Congo, Gabon, Kazakhstan, Malawi, Mongolia, Namibia, Niger, and South Africa. But the mine operators and the Yellowcake franchises are all owned by the wealthy in the world. Given that fact, it is particularly ironic that as supplies of fuel for their expensive reactors run out, the South will be last in line.

Most markedly in India, but also worldwide, this permanent menace of shortage has haunted the nuclear industry since its very earliest days. India has a critical shortage of nuclear fuel. Its only relatively abundant, but very low-grade, resources of nuclear materials are in the form of the mineral thorium. Wherever uranium does occur in India, it is feverishly extracted, even at ore concentrations a fifth or a tenth of what is considered the absolute minimum economically feasible mine grade anywhere else in the world.

Yet the industry tirelessly claims there are "many decades" of uranium fuel supply in store. The favorite number is about 60 to 100 years, which compares deliberately well with the oil industry's favorite remaining lifetime number of "about 40 years" for oil.

Again, especially in India, permanent fuel shortages require a host of "solutions," some with bizarrely dangerous and risky results. In India, as in other countries for at least 50 years, this includes the promotion of national or international programs to develop what are known as fast breeder reactors. In theory (and certainly in glossily presented PowerPoint presentations), these reactors are able to produce more nuclear fuel than they consume. Yet these doomsday machines can operate only if they have *huge* initial loads of plutonium and other high-level radioactive materials.

**AMAZING FACT: *VOLTANKEN, BITTE!***

A full-scale fast breeder reactor needs an initial fuel load of around 150 tons of plutonium, equivalent to 15,000 Hiroshima-size bombs, in radiological terms.

The risks taken with an industry-standard pressurized water reactor, itself always a potential gigantic dirty bomb, as the Japanese were cruelly reminded in 2011, pale into insignificance when compared with a large-size fast breeder reactor.

The colonial period ended in the early 1960s. From that time on, the atom has massively democratized and has been exposed to the globalizing economy. These strivings for national identity, national pride, and economic progress are quite easy to bundle together, exploit, and manipulate by industry sales teams. For the political elites, embracing nuclear power endows them with a potent symbol of technology and progress to show their nation. And symbols, despite their costs, are important for proving that yesterday's colonial underling is moving closer to catching up with its former colonial masters, and oppressors.

In India, for example, the Koodankulam nuclear plant, intended to be the largest in the country, is being built by the grace of the Russians. The chosen location is in Tamil Nadu, an ancient land home to the Tamil civilization since at least 1500 BCE, as well as the region with the highest biodiversity in this already very biodiverse country. Here elephants, tigers, and lion-tailed macaques can be found alongside some of the grandest Hindu temples, those renowned for their towering *gopurams,* or monumental stone towers. Tamil Nadu is also one of India's most urbanized states, and one of the most socially divided. At least one-third of the population lives below the poverty line. No wonder that, in the 2006 election, political promises included cheap rice at two rupees a kilogram, free gas stoves to cook on, and money for dowries. Yet Tamil Nadu's new power station costs billions of dollars. Billions! We are into fantasy money in poverty-wracked India. But fantasy is never far away in nuclear planning—it helps bridge the funding gaps.

Indians today see the growth of nuclear power as yet another proof that the country is clawing its way back to great-nation status. Sixty or seventy years ago, they say, India was an exploited colony of the British Empire. Today it is inexorably moving toward the status of an almost superpower, with nuclear clout. One member of India's Parliament well summarized the national feeling, saying:

> Of course Parliament may be divided. Political parties may be divided. But here in India the people are not divided over the issue. They are with the Government. That is the fact. We cannot bury the fact. The people of this country throughout the length and breadth of this great country are overwhelmed with a sense of pride. They are overwhelmed with a sense of joy, with a sense of

> confidence, and with the sense of pride. That is the fact. But that has not been reflected in this Parliament. At this hour, we have to stand as one. We have to express our view.

Of course, this change already can be seen in a host of growth statistics, including India's ever-rising Sensex stock index; its rate of 9 percent economic growth, with similar or higher rates applying to its electricity and coal demand; its world-class production of steel, automobiles, pharmaceuticals, and engineering goods; and its leading position in the information technology software business. Yet almost none is as important, for the power elite, as India's successful nuclear bomb tests and its ballistic missiles able to launch satellites, together with its satellite-based dual-potential range-finding and remote-sensing capabilities. And as an emerging economy, India remains low income and engaged in a race against time. Nuclear power cannot afford the consequences of cost cutting and risk taking, of constraints and shortcuts, and, least of all, of absent security features. The financing of these obligatory megaprojects, like the question of whether nuclear electricity is really cheap or not, is usually given a backseat by the political elites who choose the nuclear option. There are several reasons for this studied—or natural—ignorance of the economics, except where direct corruption plays a decisive role in winning the contract. (Not surprisingly, the roll call of potential new nuclear nations between 2010 and 2020 contains several names that figure prominently on the Transparency International blacklist—a publication of the Germany-based nongovernmental organization that monitors and publicizes corporate and political corruption in international development.)

In the new-nuclear South, the corporate strategies for adopting and acquiring the atom usually include highly complex offset and barter payment systems and methods, development aid, and increasingly carbon finance, making nuclear financing as murky and opaque as the economics of the power itself. The reality of nuclear power's intrinsic high cost and extremely high risk will certainly return to haunt the South's political and corporate deciders one day.

But there are some areas in which emerging economies in the South can have the advantage on more developed ones in the North. Savings can be made in matters such as health and safety and the environment, because, in undeveloped economies, such things are often considered to be mere luxuries that the state can legitimately decide to do away with. And it frequently does.

If nuclear power has produced a steady series of "accidents" in the developed North, how much confidence should we have in the administration of the plants in the South? In the North, each incident brings about in its wake a new safety measure; the proverbial stable door is always carefully locked after the horse has bolted. But in the South, from the very low security and safety of Indian thorium miners extracting low-quality nuclear fuels at the lowest possible price, to *Harijans* ("untouchables") carrying out the tasks of disposable industrial robots in the West, there exists in nuclear research facilities in India, as well as in other developing countries, a cultural indifference to public hazard and risk. As elsewhere but in dramatic fashion, most nuclear accidents in India are followed by a decision to "carry on as was." Nothing, even a dodgy nuclear plant, can be wasted.

In the South and across the spectrum, the environment tends to remain a "free good." Time is short, and human beings are cheaper than machines. No wonder, then, that the nuclear renaissance has rooted itself, among the dictatorships, the voiceless masses, and the corrupt and self-serving elites.

## MYTH 1

# NUCLEAR ENERGY IS THE ENERGY OF THE FUTURE

*The United States knows that peaceful power from atomic energy is no dream of the future. That capability, already proved, is here—now—today. Who can doubt, if the entire body of the world's scientists and engineers had adequate amounts of fissionable material with which to test and develop their ideas, that this capability would rapidly be transformed into universal, efficient, and economic usage?*

—The glowing promise of Eisenhower's
"Atoms for Peace" speech, 1953

*US President Eisenhower addressing the UN General Assembly in its shiny new building in New York in 1953. (US National Archives)*

**WHEN MOST OF US THINK ABOUT NUCLEAR POWER,** we picture white-coated scientists laboring in control rooms worthy of the Starship *Enterprise.* Yet when the Japanese nuclear plants started to melt down in 2011, the world was horrified to see that the scientists had disappeared and in their place were low-grade workers trying to hose seawater into the overheating reactor core.

Incongruously, since nuclear power has a 50-year pedigree in the old-nuclear countries, like Britain, France, and the United States, which makes it as much traditional as technocratic, the package still carries the slogan that it is *our future energy,* even if this claim dates from the era of black-and-white television. While everything else has changed in the economy and society, some things have not: Nuclear power has always been produced in neat but sober, huge, domed reactors and squat, windowless ancillary buildings on a distant skyline, emitting almost nothing visible, unlike those fume-belching nineteenth-century coal-fired power plants. So the story goes: nuclear power plants were always surrounded by gaunt but functional electrical transformer farms and run by calm, fatherly figures seated in front of consoles fit for *rocket science.*

But that image was finally dashed when at least four, and possibly more (for all anyone seemed to know) of the Japanese nuclear plants at Fukushima started to melt down in March 2011. Gone were the comforting, fatherly scientists, replaced by masked hardhat workers risking their lives to cool the reactors and fuel rod ponds with seawater in scenes of utter devastation. More than 150,000 Japanese were hustled out of their homes and livelihoods and moved away from the 12-mile total exclusion zone, perhaps forever.

In a single stroke, food and meat from the region became potentially radioactive. Long-lived particles, such as those of cesium-137, which can cycle through an ecosystem for decades, were scattered for hundreds of miles, ready to be taken up by plants through their roots and redeposited back into the soil when the plants rot down.

The thing about nuclear accidents—unlike, say, coal mining accidents—is that their effects are not all felt at once, and the consequences can last for centuries. In this sense, nuclear power *is* the future. But let's first step back in time, to the United Nations (UN) General Assembly of December 8, 1953. There the often-underrated Dwight D. Eisenhower, the only five-star general to become US president and an unabashed nuclear bomb enthusiast, gave what later became known as the "Atoms for Peace" speech. Speaking from the podium in Le Corbusier's modernist UN building in New York, Eisenhower offered the entire world a particularly American vision of the future, a world not only of "abundant electrical energy" but also with enough power to satisfy the needs of "agriculture, medicine, and other peaceful activities."

Actually, in those days, the UN buildings contained a grim exhibit of remnants from the cities destroyed by the first atomic bombs. A glass wall looking out toward New York City was covered in dark depictions of wailing women and crying children. And, indeed, Eisenhower's speech started with a noble apology for having unleashed the power of the atom in the first place before swiftly moving to "the future" and America's generous offer to share both the technology and the fissile material with the world—with countries prepared to accept certain rules—notably that atoms must be split solely for peaceful purposes. Doubtless the representatives of India, Israel, Pakistan, and South Africa all applauded enthusiastically at that.

Eisenhower hoped that his speech would create a huge new market for nuclear energy, led by the American firms General Electric and Westinghouse, but another goal was to distract attention from the country's ongoing series of hydrogen bomb tests, which continued smoothly following Eisenhower's presentation at the UN; the 1954 Bravo test was the United States' "biggest bang" ever. In the Pacific, Marshall Islanders still remember seeing the "second sun"—an intense fireball 1,000 times more powerful than Hiroshima—and the 20-mile-high mushroom cloud. Two years later, though, the Pacific Island tests did stop. The finale was the Flathead test of June 1956 at Bikini Atoll lagoon—a test of a dirty bomb. Because the cheapest and easiest Doomsday Machine to make is not a nuclear reactor at all but a cobalt bomb cluster. Each such gizmo is an ordinary atomic bomb encased in a jacket of cobalt. When it explodes, it spreads a huge amount of radiation. A small number of these bombs could extinguish all animal life on Earth, leaving just the insects. Undeniably, from the military perspective, nuclear power gives you real bang for your buck.

And, in other ways, Eisenhower's instincts were also right. At its heart, a nuclear reactor is a glorified steam engine, boiling water or pressurizing gas to turn turbines. Technologically, it scarcely matters that it uses radioactive metals rather than coal. It is only the financing that is complex and mysterious.

## ECCENTRIC FOUNDING HEROES OF ATOMIC MYTH

If we travel back in time to the 1930s, the founding era of atomic and nuclear science, engineering, and weapons, we will find such strange behavior—and such strange facts—that the supposedly seamless high-tech image of the atom simply collapses.

The cases of Enrico Fermi, the Italian American physicist and "godfather" both to nuclear energy and the bomb (who received the 1938 Nobel Prize in physics for identifying new elements and discovering nuclear reactions by his method of nuclear irradiation and bombardment), or theoretical chemists such as Niels Bohr and the Curies, or the Einstein and Chandrasekar Bose rivalry slash friendship, and even the real "father" of the Big Bang, Georges Lemaître, provide a strange mix of the best and worst motives. These high priests of the atom were in many cases driven by jealousy and rancor and were often simply wrong.

One example is the strange attraction and revulsion that Albert Einstein and Georges Lemaître exercised on each other, sometimes characterized as a struggle between the supposed pantheism of Einstein and the doctrinaire Catholicism of Lemaître. To some, this ideological split would explain why Einstein for many years pooh-poohed the Big Bang theory of Lemaître, which persists, relatively unchanged, today, although Lemaître called it "the hypothesis of the primeval atom."

Much more closely related to the emerging science and technology of first nuclear weapons and then nuclear power, we find that a number of the early atom scientists, such as the United States' J. Robert Oppenheimer or Germany's Werner Heisenberg, flagrantly touted the idea of atom bombs from the mid-1930s on. They did this to be heard, to get funding, and to become directors of prestigious laboratories—research labs running atom projects with an obvious military bias.

### TINKERING WITH PACKETS OF ENERGY

Tinkering with the basis of matter—that is, atomic nuclei—is the ultimate form of tinkering. It underlines a central fact: The antique version of nuclear science lurking behind modern atomic energy is much less complicated than the way it has been presented and packaged by the nuclear industry.

The easily grasped idea of the planetary atom, such as the Sun and its planets, dates from the late nineteenth century. The concept was just beginning to fall apart in the 1930s era of the atom scientists. After that time, nuclear science drifted in one direction only: toward abstraction and complexity. Today we can call atomic and nuclear science the study of statistical packets of energy. Not so much a science of "things" but rather, conjuring performed with mathematics.

Mirroring this process today, nuclear finance has become so unremittingly complex and mysterious that few can understand it. The numbers are so large, and the underlying assumptions are so strange, that the mind balks. However, just as with the early atom machines of the 1940s and 1950s, there is a simple explanation. *Dangerous tinkering* is the very same key to understanding the illogical and fragile world of nuclear finance.

One of these, from the early 1940s, was Enrico Fermi's Manhattan Project. Its aim, of course, was to develop weapons of mass destruction—monstrous weapons to kill civilians, young children, elderly grandparents, cats and dogs—all indiscriminately, all uncaringly, unthinkingly even. But not through the exploding of nuclear reactors, mind you. These lethal devices were just a by-product.

One might ask why, if nuclear power is the energy of the future, it has taken so long for it to supply less than 6 percent of the world's energy. Even this meager achievement accounts only for commercial energy, excluding the requirements of world total primary energy of all kinds. In that case, nuclear power's share drops even further, to about 2.5 percent.

### JARGON BUSTER: PRIMARY AND SECONDARY ENERGY

The key point about world energy is that it is almost all thermal. Whether it is created by burning coal, oil or gas, or firewood or dung, or even running nuclear power plants, the first thing produced is heat. This heat is what is defined as primary energy, and for nuclear plants, getting economic value out of primary energy almost invariably involves upgrading it to electricity—energy in a form people can use more easily and thus be made to pay for.

Heat energy from coal can be used directly to run stoves in homes, smelt iron in factories, or even to toast muffins. The same flexibility applies to gas. However, the heat generated by nuclear power rarely has any direct and immediate commercial value. Only when it is upgraded or converted to electricity is it viable.

Chances are that Japan's nuclear disaster—which occurred in reactors designed in the 1960s (one of which was only days from its fortieth birthday)—will actually shift the nuclear industry backward. Modern atomic and nuclear physics has a term for this, originally coined to describe black holes: the "event horizon," a context from where no return is possible, a time, matter, and energy nexus where the outcome cannot replicate the start of the process. In the words of Lemaître, explaining that evocative primeval atom theory: "A tomorrow for which there is no yesterday."

### THE NUCLEAR POWERS

Despite its nuclear power ambitions, as of 2010, India falls just outside the top 12, at number 14, just behind Taiwan.

| Country Δ = also member of the nuclear club | Nuclear Operating Capacity[1] (in gigawatts) | % of National Electricity[2] | % of National Primary Energy[3] | % of National Energy[4] |
|---|---|---|---|---|
| 1. United States Δ | 101 | 20 | 9 | 4 |
| 2. France Δ | 63 | 76 | 38 | 15 |
| 3. Japan | 47 | 25 | 13 | 5 |
| 4. Russia Δ | 22 | 17 | 6 | 3 |
| 5. Germany | 20 | 28 | 11 | 6 |
| 6. South Korea | 18 | 36 | 14 | 7 |
| 7. Ukraine[5] | 13 | 47 | 17 | 9 |
| 8. Canada | 12 | 15 | 6 | 3 |
| 9. United Kingdom Δ | 10 | 13 | 8 | 2.5 |
| 10. Sweden | 9 | 42 | 28 | 8 |
| 11. China Δ | 8 | 2 | <1 | 0.4 |
| 12. Spain | 7 | 18 | 9 | 4 |

1. Nuclear capacity in operation and under construction as of January 2010, Public Services International Research Unit figures. The total world installed capacity was 375 gigawatts that year.

2. 2008 International Atomic Energy Authority figures.

3. As of 2009, according to the *Statistical Review of World Energy.* Typically, electrical energy is about half of all primary, or commercially tradable, energy a country consumes.

4. Assumptions about national energy vary widely. For example, in France, a significant proportion of home heating is by locally chopped wood sold on the black market, but little of this appears in energy statistics. In the table, it is *conservatively* assumed that electricity provides one-fifth of total energy, not just so-called primary (commercial) energy.

5. Lucky old Ukraine inherited many of the former Soviet Union's power stations, including Chernobyl.

Today, for the nuclear industry, an event horizon is emerging. The industry is facing economic, financial, uranium resource, technology, environment, and public awareness, and other forces and factors that could generate a nuclear winter, after which no recovery will be possible.

The Fukushima disaster, which quickly reached level 7 on the international scale for nuclear accidents (the highest possible level, the same as Chernobyl), was an old-style atomic industrial disaster, similar to a long but little-known string of disasters stretching back to the dawn of the atomic age.

And the cost of the Japanese nuclear setback will likely be as high or higher than Chernobyl, even though the immediate death toll has been lower. Taking the real economic damage from Chernobyl as approaching the equivalent of $250 billion in today's money, this incredible sum is still a mere fraction of what it might have been had the radiation from the plant not sterilized for decades, and perhaps centuries, a region of relatively sparse population and low-value agricultural land, in its 3,000-square-mile total exclusion zone. Such luck does not apply to the unfortunate Japanese; the Fukushima site is in much more densely populated northeast Japan, in valuable agricultural land. Its location approximates much more closely the pattern, distribution, and density of nuclear power plants in most countries.

Environmentalists and health specialists will argue for decades about the significance and effects of this latest nuclear accident. However, economists and politicians, including many who previously supported nuclear energy, are in agreement on one thing: For the Japanese and for the nuclear industry as a whole, the Fukushima incident is an unvarnished, unmitigated economic disaster.

Yet the specialist journals and the Sunday papers still insist that nuclear energy has a bright future. Oil is running out, coal is too dirty, renewables are simply irrelevant. Nuclear, they insist, powers the world's major economies and is poised to power the emerging world too.

We have been long told that, compared to the depleting, outdated, polluting, and geopolitically controversial resources such as oil, natural gas, and coal—or the expensive and difficult-to-exploit renewable energy sources such as wind and solar—nuclear energy is the perfect solution. There's a nearly unlimited supply; it is clean and green; and it is cheap—so what is not to like? Opponents to nuclear power seem almost to share the delusions of King Canute warning the tide to turn back.

Take the world's biggest economy, the United States. Potentially, the United States can produce and enjoy 100,000 megawatts (MW) of nuclear electricity each year—far more than any other country. That is nearly enough for each of the country's 120 million or so homes to simultaneously boil a kettle and watch television. (But there will be no energy left to take the subway to the office, which would be unlit, unheated, and shut anyway.*) Similarly, sometimes we are impressed to read that nuclear power is the key element in the French economy, providing nearly 80 percent of French electricity, and that in the Far East, China is racing to install new reactors to power its new cities. In neighboring Japan, the share of nuclear power "in the national energy portfolio" is 30 percent, as Nassrine Azimi wrote recently in an authoritative opinion piece for the *New York Times* that put into context the "existential moment" that was Fukushima. Plus nuclear is poised to provide India with a quarter of all its electricity within, say, the next ten years. No wonder that even the oil-exporting powers of the Middle East have got the nuclear bug. The United Arab Emirates has a plan to build four new reactors to generate "about a quarter of the Emirates' power by 2020"; Iran and Iraq have built small reactors; and Saudi Arabia is planning a nuclear-powered city. A nuclear-powered city! That beats those eco-villages with grass on the roofs.

Yet this picture is much more myth than reality. Sprinkle it with just a few plain facts, and it quickly dissolves.

---

* Actually, supporters of nuclear power would dispute this fact. They dream not only of nuclear-powered trains but even of nuclear-powered cars! But the essential point remains that, even if nuclear investment focuses on the most efficient, large-scale generators of power, those 1,000-MW power stations, the total energy generated will meet only a relatively trivial amount of today's energy requirements.

**AMAZING FACT: SAVE ENERGY!**

- Despite the huge fleet of reactors in the United States—twice the number of its nearest competitor—nuclear generation provides only about 5 percent of the country's *energy.** It provides a much more important-sounding proportion of the electricity, but that is neither here nor there. Industries (or even home heating systems) at present burning coal, oil, and gas—the fossil fuels that make up about 85 percent of America's energy mix—could as easily (if not as economically) use electricity, and vice versa. And of course, simple energy conservation measures could immediately save many times that amount of energy.
- In France, even after 35 years of nuclear power development, the "nuclear dreamland" still gets only one-sixth—about 16 percent—of its final energy from nuclear power. The reality today is that almost half of France's *energy* consumption still comes from oil (despite the nation having almost no oil reserves of its own).
- India's expectation that nuclear power will supply a quarter of its electricity by 2050 seems more than a little optimistic, as at present nuclear power supplies a mere 2 percent of the country's electricity, which equates to a pitiful near-zero percentage of its *energy*—despite the fact that 20 reactors already are humming. What is really going on in India, a land where most people are still not connected to the electricity grid (what grid?), is that large amounts of money are being directed toward the business interests of a new, nuclear elite.
- As for Japan, the *New York Times,* normally so fastidious in its attention to facts and subsequent "corrections," is simply wrong. Japan gets a mere 5 percent of its *energy* from nuclear power—although, prior to Fukushima, it did generate almost a quarter of its electricity that way. Funny how the distinction between a country's energy needs and its electricity provision keeps getting confused—always in nuclear's favor! Curiously, following Fukushima and the forced shutdown of a quarter of Japan's nuclear power stations, a series of rolling blackouts affected Japanese cities. But consumers were

* Some experts put it higher—but not by much: 8 percent. The US Energy Information Administration gave this higher figure, along with an 85 percent figure for the contribution of fossil fuels, in a report for 1998; the proportions have not shifted much since then.

> right to be puzzled by the shortfall. The excess electricity-generating capacity available to the national grid should have more than made up for the lost output. Could it be, some asked, that the nuclear utilities *wanted* to make their role appear more significant than it really was?*

Of course, national "energy" statistics are extremely imprecise, due to immeasurables such as biomass, which is still by far the most important form of energy for the majority of the world's population. (This is the measurement of people cooking food and heating their homes with wood, for example. Even nuclear-powered France relies on huge amounts of uncounted locally cropped logs for heating.) Again, industrial plants generating their own energy privately are rarely factored in. Journalists seem too busy to sort out the wheat from the chaff, let alone the national energy requirements from offers by huge companies to generate more electricity. In examining energy issues, it pays to be skeptical. After all, energy is what the universe is ultimately made of, and it is certainly what drives economies.

### AMAZING FACT: IS NUCLEAR THE 2.5 PERCENT SOLUTION?

- In 2009, the world generated 2,558 terawatt-hours (TWh) of nuclear electricity, about 13 percent of the world's commercial electricity. (A terawatt is 1 billion kW.)
- After 50 years of promising to replace fossil fuels, nuclear power provides only about 6 percent of world primary or *commercial energy* and around 2.5 percent or so of the total world energy requirements of all kinds.

---

* In fact, contradictory though it might seem, both the skeptics and the authorities could be right. Although in theory the Fukushima losses should have been possible to cope with, the earthquake plus heightened safety concerns led to the shutting down of several other nuclear plants, meaning that out of TEPCO's (the Japanese electricity utility) *theoretical* nuclear power capacity of 18.4 MW, it lost 10.2 MW on March 11 and was left with less than 5 MW of *actual* capacity.

Yet, as with most technological solutions, novelty has shielded nuclear from a clear-eyed assessment of its actual usefulness. Many people in the "old-nuclear" countries, where civil nuclear power was first developed from the 1950s and 1960s, can still recall how atomic energy was first presented to them. In schoolbooks, in speeches from leading politicians, and in the press, an impressive chorus of recommendations for ever bigger and bigger high-tech reactor projects presented nuclear power unambiguously as the energy of the future.

However, the same people may also remember how nuclear power suddenly faded out in the late 1970s, especially after 1979. This was the year of Three Mile Island in the United States, the first major civil nuclear accident. Or at least it was the first one to receive massive media exposure. Following this, for the industry, came the so-called nuclear winter.

During this winter, which lasted over 20 years, reactor orders and completions dried up. Abandoned by corporate elites, nuclear power was also shunned by politicians, except in a few sturdy cases, such as France. Yet, just seven years later, the Mitterrand government tried to wish away the deadly cloud of radioactive fallout from the Chernobyl disaster that settled on rural France like an invisible shroud. But then, let's not forget, France is the world's most dependent nuclear addict, with almost 80 percent of its electricity produced in atomic reactors.

And then, with a new millennium starting, nuclear power was back—suddenly reinvigorated with a surge of multibillion-dollar projects—or at least plans for them. Once again, the atom was to be the energy of the future.

That is why, when proponents of nuclear energy came to present their 2050 Roadmap for Nuclear to the World Nuclear Association at a conference in 2010 (just prior to the Japanese accident), they were hoping that rules governing new reactors might soon be slackened a bit, arguing in the name of "harmonization" that the eventual aim should be for national regulators to "accept the conclusions of design reviews conducted by other regulators without having to duplicate the work themselves." In other words, US reactors could conceivably be built to Chinese standards.

**NUCLEAR ADDICTS: THE MOST DEPENDENT OF THE 12 BIG USERS**

| Country<br>Δ = also member of the nuclear club | % of National Electricity from Nuclear[1] |
|---|---|
| 1. France Δ | 76 |
| 2. Ukraine | 47 |
| 3. Sweden | 42 |
| 4. South Korea | 36 |
| 5. Germany | 28 |
| 6. Japan | 25 |
| 7. United States Δ | 20 |
| 8. Spain | 18 |
| 9. Russia Δ | 17 |
| 10. Canada | 15 |
| 11. UK Δ | 13 |
| 12. China Δ | 2 |

1. 2008 International Atomic Energy Authority figures. Despite its grandiose nuclear power ambitions, as of 2010, India was not a big user, falling just outside the top 12, at number 14 (just behind Taiwan). Indeed, China only just scrapes in.

Once the safety issues had been dealt with, the industry experts had solutions for the finance problems too. "Governments should consider some form of support or guarantee for private sector investment in new nuclear plants," where the "risk/reward ratio," as they put it, would otherwise deter investors. How much money would they like to see put at the disposal of the world nuclear industry? About $4 trillion—four thousand thousand million dollars.

Eisenhower promised at the very least that nuclear electricity would be "economic" when he was not talking about it being "too cheap to meter." Yet the bottom line is that when Eisenhower made his speech, *all* forms of electricity were relatively expensive, whereas nowadays it is ridiculously cheap. In the United States today, off-peak electricity sells at

**NUCLEAR ADDICTS: THE MOST DEPENDENT OF ALL THE NUCLEAR USERS**

| Country<br>Δ = also member of the nuclear club | % of National Electricity from Nuclear[1] |
|---|---|
| 1. France Δ | 76 |
| 2. Slovak Republic | 56 |
| 3. Belgium | 54 |
| 4. Ukraine | 47 |
| 5. Sweden | 42 |
| 6. Slovenia | 42 |
| 7. Switzerland | 39 |
| 8. Armenia | 39 |
| 9. Hungary | 37 |
| 10. South Korea | 36 |
| 11. Bulgaria | 33 |
| 12. Czech Republic | 32 |

a shade under $40 per megawatt-hour (MWh), or 4 cents a kW-hour, a sum that the federal government tops up the providers with by another $18 per MWh. (These were the forward prices for 2012.) In the United Kingdom and in France the nuclear industries are already cosseted with politically determined rates for wholesale electricity, a comfy near £50/MWh and €55 odd/MWh respectively; call it $80/MWh.

Yet despite that huge hike in energy prices courtesy of the Europeans, and despite increasingly frenzied efforts to reinvigorate the reactor market (usually by producing newer and newer reactor designs, presented as new generations, that come and go without a single plant actually being built), ever since Fukushima, the antiquated, outdated, dangerous, costly, and badly organized reality of nuclear power has come to the fore. The situation is comparable to that which has occurred in some other industries and other lost causes. Whether we are talking about rearguard action to keep the asbestos or polychlorinated biphenyl industries alive, to keep producing DDT, to save the Concorde jetliner, to go on using videotape recorders, or to not abandon the US space shuttle, in the end, attempts by private, corporate, and state interests to maintain demonstrably dangerous, useless, costly, or outdated industries cease only if and when the facts become widely known.

Sometimes that process is very slow, until some trigger event takes place in a context where there is rising public concern and increasingly desperate action by the corporate and political elites who defend the failed technology or industry. At the time, the trigger events may not be recognized for what they were.

For nuclear power, these trigger events have been rapidly accelerating. Perennial pressures on the industry range from the financial, economic, and industrial to massive national security risks, the radiological implications of nuclear power, and the danger of atomic materials and waste for human health and the future of life on the planet. Add in serial disasters, and the outcry for abandonment picks up speed.

Today, several indicators show we are likely near to one-way and massive change, owing to the radical upturn of public awareness regarding the extreme dangers of nuclear power. Like a Russian doll, the nuclear system is highly compartmentalized, has always been dark and secretive, and contains a potentially massive series of shocks in reserve. For its critics, the countdown to nuclear power's event horizon is already under way.

Mind you, the nuclear industry has survived previous existential crises; most notably, and inexplicably, the problem of its escalating costs. In this case, it grasped a green branch that arrived just in time to save it from the environmentalists who for so long were its enemies.

Ron Cameron and Martin Taylor, analysts at the Nuclear Development Division of the Organization for Economic Cooperation and Development Nuclear Energy Agency in Paris, sum up the thinking that the industry now clings to more desperately than ever:

Without decisive action, energy-related emissions of carbon dioxide ($CO_2$) will more than double by 2050 and increased oil demand will heighten concerns over the security of supplies. To change our current path will take *an energy revolution* and low-carbon energy technologies will have a crucial role to play. Nuclear power among other technologies needs to be widely deployed if we are to reach our greenhouse gas emissions goals.

# MYTH 2

# NUCLEAR POWER IS GREEN

*Greenhouse gas emissions, if continued at the present massive scale, will yield consequences that are—quite literally—apocalyptic.... If these predictions hold true, the combined effect would be the death of not just millions but of billions of people—and the destruction of much of civilisation on all continents.*

—John Ritch, director general, World Nuclear Association

*Not merely greenhouse gases, but clouds of radiation being poured into the atmosphere from the burning core of the Chernobyl nuclear power station, in May 1986. (AP Photo/Igor Kostin)*

**ONCE UPON A TIME, ALL SELF-RESPECTING ENVIRON-**mentalists hated nuclear power. It produced invisible pollution—radiation—that seeps everywhere, causing genetic diseases that interfere with nature. It left toxic residues that were poisonous for thousands of years. It was statist, secretive, large-scale, high-tech, complex, and expensive—the very antithesis of virtuous, simple ideals of many environmentalists. In all these fundamentals, nuclear energy fundamentally conflicts with the core green values of simplicity, small-scale solutions to local needs, and zero pollution.

And then along came the destabilizing idea that carbon dioxide ($CO_2$)—indeed, carbon, the fundamental building block of all animal, plant, and human life on Earth—was the *real* pollutant to worry about. Almost overnight, this new threat reversed green politics, as world governments suddenly were called on to seek high-tech solutions to fossil fuel dependency.

If the political "consensus" about climate science today has collapsed, with not only China and Russia but Canada and Germany stepping up use of fossil fuels, it lasted long enough for there to have been one important outcome. After the virtuous Kyoto Protocol was negotiated in the mid-1990s, countries that had turned away from nuclear power (due to its unsolved waste problems and the ongoing risk of nuclear meltdown) embraced it again, specifically citing the danger to the planet from greenhouse warming. Even eastern European countries, those literally in the shadow of Chernobyl, signed up for replacements for their old Soviet atomic reactors worth a good $50 billion.

In the rush away from dirty, dangerous carbon, nuclear energy made a miraculous return to favor. That is why, as early as 2001, the International Energy Agency announced that climate change had altered the future for nuclear energy, and why, in 2010, the British Royal Academy of Engineering, representing contractors involved in numerous nuclear power projects around the world, was confident enough to ask, in a pamphlet titled *Nuclear Lessons Learned,* "Does the Government need to do more to ensure investors select low-carbon options for future electricity generation?"

In the early days of nuclear power, Canada and Sweden were two key backers: Publicly, they are countries with a great sense of social and environmental responsibility and worries about coal-produced acid rain; privately, they are countries with a deep concern for their own struggling nuclear industries. Similarly, when today Australia and the United Kingdom respectively appear to combat global warming through a mix of energy taxes and energy handouts, it is industry lobbyists, not environmentalists, who are driving the decisions. In the United Kingdom, nuclear power is still the energy of choice for governments of both right and left, while in Australia, exports of gas, uranium, steel, and other raw materials are far more lucrative than digging up coal.

In fact, the United Nations' newly dreamed-up International Panel on Climate Change (IPCC) was pronuclear from the start. Its first chief, a Swede, Bert Bolin, was active in Swedish energy politics, which relies on the two pillars of hydroelectricity and nuclear power, and was admired for forcing the German government to install expensive sulfur filters on its coal-fired power stations (ostensibly to reduce acid rain in Sweden). The second IPCC head, Robert Watson, was a research director at the World Bank with a reputation for actively promoting dams in the Amazon rain forest and nuclear energy for everyone else. A third key climate change activist, a German named Wolf Häfele, not only invented the 20 percent figure for $CO_2$ emission reductions that became the go-to figure for climate change politics for two decades (after it was made official by the final statement of the Toronto Conference "Our Changing Atmosphere" in 1988) but was a key player in the development of a new type of nuclear reactor, the dramatically expensive fast breeder. According to the energy policy analysts Aynsley Kellow and

Sonja Boehmer-Christiansen, Wolf arrived at his 20 percent target by a peculiar route. Other activists might have tried for a higher number—perhaps 60 percent—but Wolf argued that the nuclear technology was not ready yet. He showed splendid aplomb, given that the debate was taking place in the shadow of the 1986 Chernobyl nuclear disaster that had spread a cloud of poisonous radiation over much of Europe. Yet, just two years later in Sweden, climate change was cited as the reason *not* to phase out nuclear power in the country.

### EVERY CLOUD HAS A SILVER LINING

When Chernobyl melted down, spreading radiation that is estimated, albeit controversially, to be responsible for the early deaths of 100,000 people, it certainly looked like bad news for the nuclear industry. Yet every cloud has a silver lining, and so it proved here. By international treaty, nuclear companies have only very limited liabilities. So it is that two decades later, an unprecedented *international* effort, equal in its own way to the entire "climate science" research effort, has directed some $2.5 billion into cleaning up after and making safe the damaged reactor. Put another way, the profits to be made from cleaning up after exploding nuclear reactors far exceed those to be made running them safely. And is the industry ashamed to benefit from its own disasters? Not at all; these days the Russian firm behind the design even has the chutzpah to use Chernobyl in its advertising for "extra-safe reactors."

Even if global warming science was not explicitly invented by the nuclear lobby, the science could hardly suit the lobby better. It seems to require directing previously unimaginable subsidies toward "carbon-free" energy, of which nuclear claims to be the only serious option at present, a trick achieved by skewing the debate toward electricity production and ignoring the oldest forms of energy—wood, water, and animal dung—which still are key energy sources for many people—and not only in the developing world. Once the nuclear industry convinces everyone it is the *key* energy source, rather than the irrelevant and highly costly cherry on the pie, its fossilized competitors are subject to crippling costs.

Thus it was on December 10, 2009, that an extraordinary group of 100 world leaders and their most esteemed scientific advisors gathered in gloomy Copenhagen, Denmark, for an unprecedented conference with just one task—but a huge one: saving the planet. The lead item on the agenda: how to stop runaway global warming.

The 15,000 delegates and officials, 5,000 journalists, assorted heads of state, plus celebrities including Leonardo DiCaprio, Daryl Hannah, Helena Christensen, George Clooney, Archbishop Desmond Tutu, and Prince Charles, were in Copenhagen to decide how best to reduce emissions of a deadly gas—$CO_2$—that was already thought to have caused droughts, the melting of the polar ice caps, and deforestation and the spread of diseases in the tropics.

Nobel Prize winner and senior US statesman Al Gore had set the tone earlier, when he declared in his popular documentary film *An Inconvenient Truth:*

> Humanity is sitting on a time bomb. If the vast majority of the world's scientists are right, we have just ten years to avert a major catastrophe that could send our entire planet's climate system into a tail-spin of epic destruction involving extreme weather, floods, droughts, epidemics and killer heat waves beyond anything we have ever experienced—a catastrophe of our own making.

Yet a week later, the conference broke up and the 15,000 delegates, 1,200 limos, and 140 private planes departed, in a swish of $CO_2$ emissions, with nothing decided: no $CO_2$ reduction, no emergency aid for threatened countries, only a vague promise to look at the issue again . . . in five years.

What went wrong? It seemed as if the entire climate change campaign, which had dominated national and international politics for 20 years, ever since the Kyoto Protocol was signed in sunny Japan, had fizzled out with hardly a whimper. Or, more accurately, that climate change policy was a house of cards that had collapsed with the first breath of cold wind.

That Kyoto Protocol had been a remarkable achievement. On its face, it committed every country that signed up to reduce $CO_2$ emissions.* Actually, in fact, it did not do so, as, by a rather neat sleight of hand, "target levels" were set for reducing emissions that countries were already well under. In reality, Kyoto was less about reducing carbon dioxide emissions and more about countries agreeing among themselves to pay more for their electricity supplies. Or rather, every government that signed agreed to make consumers pay more. Hapless energy users the world over would pay about $350 billion, in return for which global temperatures were supposed to be reduced by *about one-fifth of one degree.*

To put that in perspective, the same amount of money could have satisfied, at a single stroke, access to basic healthcare, education, clean water, and sanitation for all Third World inhabitants. It is true that such things are not useful if you happen to have starved to death by desertification in Africa or been submerged under rising seawaters in the Pacific—if that is really the choice. But to swap these options for a drop in temperatures of one-fifth of one degree?

You would have to be very green—that is, naive—to sign up for that. Or very cynical. And, of course, there are plenty of both kinds of people.

Originally the role of $CO_2$ in the atmosphere must have seemed like a very good subject for scientific research and academic debate. But somewhere along the line it also became a tool for governments to intimidate their populations into passive acceptance of very real changes—from the tiny (switching to fluorescent light instead of incandescent; permitting wind turbines to crowd previously sacrosanct hilltops) to the major, such as widespread destruction of rain forests for biofuels and accepting nuclear power plants and all their dangerous consequences.

The counties pressing the climate change arguments during the Kyoto conference had in their national interests an array of businesses linked to

* Despite making supportive noises, the United States never, in fact, ratified the treaty.

**INGREDIENTS THAT MAKE UP THE WORLD ENERGY PIE**

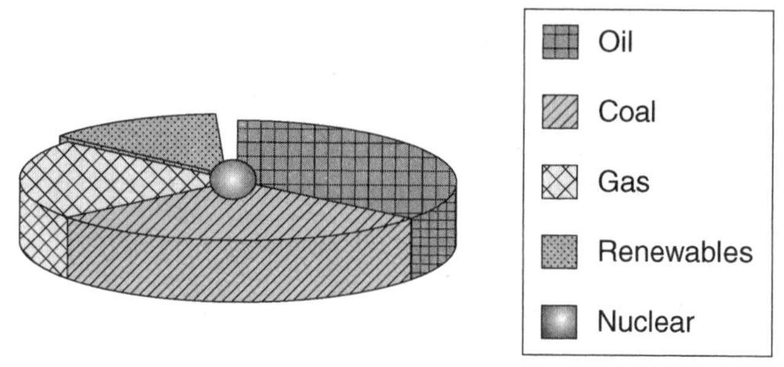


This is the primary energy pie, consisting of both commercial, tradable energy and non-commercial, informal energy sources, such as wood for heating.

*Notes:* Oil provides 48,000 terawatt-hours (TWhs); 1 million tons oil = 4.5 TWhs.

What is a TWh? One TWh corresponds to $10^{12}$ watt-hours, or the energy required to heat approximately 50,000 houses in advanced industrial countries for one full year.

Coal provides 39,000 TWhs; gas, 30,000 TWhs; and renewables, 18,000 TWh. Renewables other than hydroelectricity include wood, both commercial and noncommercial, animal dung, solar heat and electricity, wind electricity, geothermal energy, nonwood/nondung biofuels, and others. Wood and dung are major sources of energy, along with hydroelectric power, which on its own contributes some 3,000 TWhs of energy, approximately 2.5 percent of the total world energy mix.

Finally, nuclear power provides just 8,000 TWhs. In 2010, nuclear secondary output—electricity—equaled about 2,900 TWhs. That makes it the cherry on the world energy pie.

*Data Sources:* IEA, US EIA, BP Statistical Review

and dependent on nuclear power as well as concerns about the high cost of domestic subsidies for coal. How many times have we been told that oil money is behind climate skepticism? That is true only if it is a kind of cunning double game. Oil companies are also one of the winners in the drive against coal, and restrictions on mining make their gas holdings more valuable. And all governments stand to gain from policies that result in a greatly expanded tax base, particularly if the new taxes are on essentials, such as energy, and difficult to evade. Many of the protocol's supporters—Germany, Japan, and the United Kingdom—had very expensively subsided (and unionized) coal industries of their own that their governments would have liked to run down. By the end of the twentieth century, for example, Japan was paying five times more to mine its own coal than it would have cost to buy coal on the open market from Australia.

For this reason, climate change was originally, and remains, a rich country's hobby dominated by a club of largely Anglophone countries: The United States, the United Kingdom, and Canada lead the research, supported by those reliable allies, Australia, Germany, Japan, and the country with the original (acid rain) interest in atmospheric pollution: Sweden.

Because of these national interests, climate change became a simple story about too much $CO_2$ directly pushing up global temperatures. The politics behind it, however, is far from simple and anything but progressive. As mentioned, one factor was the political drive to shut down the highly subsidized coal industries of Germany and the United Kingdom. Another was the realization that a protocol that made nuclear power "clean" would be immensely profitable to certain advanced economies. But a third was the realization that the creation of a new trade in licenses to emit $CO_2$ could be the most lucrative game ever, potentially generating trillions of dollars. The sums involved in energy policy are enormous.

Money like that was enough to get many scientists to agree with the alarming research on climate change. But to collect all the voices of the on-message scientists and to make sure they were heard, the "rich countries club" created the IPCC in 1988. The organization was based on the United Nations (UN) structure that had been proved so effective in getting worldwide assent to environmental regulations controlling acid rain and gases suspected of causing holes in the ozone layer and of course, public consent to nuclear power, via the International Atomic Energy Authority.

If the book-length research reports of the IPCC are scientific, few of the resulting key summaries for policy makers are; rather, these papers are edited line by line by political appointees of national governments.

Many of the latter are hard-nosed political animals from Sweden, Canada, and the United States. They include Al Gore and his acolyte, Timothy Wirth, who as undersecretary of state for global affairs once stated: "We've got to ride this global warming issue. Even if the theory of global warming is wrong, we will be doing the right thing, in terms of economic policy."

This cynic wrote this famous line in the UN's landmark report on global warming (the Second Assessment report of the IPCC): "The balance of evidence suggests a discernible human influence on global climate."

Being political in origin, the output of the IPCC's research is curiously immune to what Karl Popper, the twentieth-century philosopher of science, calls "falsification." Most claims are couched in terms of "probabilities" with such wide error bands that future events could scarcely fall outside them. Indeed, some of the few sections that could be falsified have caused embarrassment later—like the claim that the Himalayas would melt "by 2035" or that, in a hotter world, malaria would extend its range to Europe. (Both claims are ill-informed. It would take many centuries for the Himalayas to melt even in an incredibly hot world; as for malaria, it was endemic in Europe for centuries—but then, rather than the weather changing, the marshes were drained.) The nice thing about climate change for politicians, though, is that for many of the claims, just like a politician's answer, there is no possibility of falsification. There is nothing that counts as evidence against. Increased rainfall in the northern hemisphere is evidence of climate change, but so too is decreased rainfall in the southern hemisphere. Or vice versa, when it suits. Melting of ice in the Arctic is evidence of climate change, but the observed cooling of the Antarctic is not problematic: It can be explained in terms of "other effects," such as changes in ocean currents. Hot summers in the one place are excellent supporting evidence, but cooler summers in another (e.g., in northern Europe) are not disproofs but merely require additional ad hoc hypotheses.

Today no one thinks that human emissions of $CO_2$ are going to be reduced, let alone phased out, as the original policy required. If $CO_2$ really causes the planet to overheat, then, assuredly, it *is* going to overheat. The subject is essentially a scientific debate. It therefore has its shades and nuances of meaning but contains no neutral truths and remains far from settled. Not so the political debate. This debate is—and has always been—both unsubtle and unequivocal. For politicians around the world, climate change has provided a useful stick with which to beat opponents. It represents the triumph of opportunism over principle and of special interests over social justice. That is why the nuclear industry has flourished in its shadow.

With the paid-for support of both the scientific community and the free but erratic and ill-informed green campaigners, it was open season on "dirty" fuels, such as coal and oil, with money all around for new initiatives—hydroelectric projects and biofuel plantations in the tropics, not to mention considerable amounts for the plucky, if negligible, contributions of solar and wind. But the biggest winner in the climate change drama has undoubtedly been nuclear power.

Enter the Public Intellectual Greens, to be distinguished from the public intellectual environmentalists, whose fault, if they have one, is merely being a bit smug. Once the apparatchiks of the UN and international politics had done all they could to create the new science of climate change, the next job was to win over public opinion—to convert the taxpayers and consumers. And it was here that the role of intellectuals became vital. These are people such as Stewart Brand in the United States, Bruno Comby in France, and James Lovelock and George Monbiot in Britain, all well-known authors as well as (respectively) an electrical engineer, an inventor, and an environmental columnist for the *Guardian* newspaper in London.

Comby is well-known in green circles, but he started life as an electrical engineer for France's nuclear industry, before a change of career path led to a book called *Delicious Insects* in 1990. Six years later, Comby set up the Association of Ecologists for Nuclear Power (AEPN is the French acronym), which claims 10,000 members worldwide and says that nuclear power is, well, "the future." The apparently independent work of such ecologists underpins the campaigns of the international stars of climate change doom, led by Al Gore and Rajendra Pachauri, as well as a galaxy of professional green campaigners, such as Patrick Moore. The cofounder of Greenpeace International, Moore was one of the "rainbow warriors" who narrowly avoided being blown up by the French secret service on the boat of the same name and now runs a pronuclear institute called the Clean and Safe Energy Coalition. These "ecologists" want, in particular, to see nuclear power exported around the world, saying that, in their considered view, it is the only "clean" way to generate power.

The influence of some Public Intellectual Greens is considerable. In France in 2011, for example, Nicolas Hulot tried to become the ecologist party's presidential candidate and indeed, earlier on, exercised a sort of political power by fronting a whole range of energy taxes solely motivated by concern about climate change. Although his personal ambitions came to naught, in the (post-Fukushima) 2012 French presidential election, the official line of the Greens there was to maintain the existing number of nuclear power stations until new renewable sources might begin to make them obsolete.

Another influential environmentalist is Stewart Brand, famous in the 1980s for his *Whole Earth Catalog* of green goodies, who in recent years has recommended a diet of small nuclear reactors, pointing out (correctly) that windmills and solar panels produce only negligible amounts of electricity. Acknowledging that this advice remains controversial, he told a newspaper that he was not trying to be pronuclear, only "pro-arithmetic." Jean-Marc Jancovici in France has echoed that sentiment, pointing out that since today fossil fuels supply 80 percent of the world's energy, the question has to be asked, What, once they are exhausted, is going to replace them? This is indeed a good question, but Jancovici's reply, coupled with assurances that nuclear waste is less hazardous than pesticides and "does not explode," is more illustrative of the rhetorical style of green campaigns than actual, facts-based arguments.

The eccentric and aggressive Monbiot advises true environmentalists to join with him in advocating nuclear power as the second-best way to "decarbonize the electricity supply." His first choice, oddly enough, is gas with carbon capture and storage. This solution holds up even less well under scrutiny. Suppose for one moment his plan was adopted—how long before acceptably "clean" gas reserves would be exhausted? Certainly it is not long enough to provide any sort of a solution to future world energy needs.

Monbiot is, however, only a lukewarm convert to nuclear; hence his gas dream. Better to listen to the argument of another British self-styled environmental warrior, Mark Lynas, who, when he is not campaigning for nuclear power, can be found hunched under oilskin in the Antarctic or sweating in a jeep in the Amazon rain forest. In 2008, Lynas set out all the reasons he could think of for supporting nuclear power in an article for the left-leaning British political magazine the *New Statesman,* called, practically enough, "Why Greens Must Learn to Love Nuclear Power." He swiftly dismisses the problem of nuclear waste by citing a study that says most of it will decay away naturally in "less than a thousand years."

Lynas also trots out the low estimates of deaths from nuclear accidents past, using such sources as the US Nuclear Regulatory Commission (a government body, with a structural interest in minimizing the consequences of nuclear accidents) and a book by Professor David MacKay called *Sustainable Energy—Without the Hot Air,* which attempts to prove that nuclear power is one of the safest ways to generate electricity, with only about one death for every 10 gigawatts a year. But the apex of Lynas's argument is that the IPCC (yes, that group again) says that nuclear produces hardly any $CO_2$ emissions—about "40g $CO_2$-equivalent per kilowatt-hour" (whatever that may mean), which is similar to good old wind power or "renewable electricity from other countries' primarily from solar farms in the North African desert." Naturally, Lynas prefers wind turbines, but he adds: "[I]t is vital to stress that neither I nor MacKay nor any credible expert suggests a choice between renewables and nuclear: the sensible conclusion is that we need both, soon, and on a large scale if we are to phase out coal and other fossil fuels as rapidly as the climate needs."

So what is the true Green response to world energy needs? Lynas notes that "an anti-nuclear report" argued that an additional 2,500 reactors would need to be built by 2075 in order to significantly mitigate global warming and that the report's authors seemed to have thought that this was a pipe dream. On the contrary, he says, "it sounds eminently achievable," since it is, he calculates triumphantly, "only a five-times increase from today."

And finally, on to a much grander if highly eccentric figure, former NASA scientist James Lovelock, a key proponent of both catastrophic climate change theory and salvation by nuclear power. Lovelock is typical of the newest and most schizophrenic thinking, where doom is constantly threatened unless we learn to love the friendly atom. He says, "What makes global warming so serious and so urgent is that the great Earth system, Gaia, is trapped in a vicious circle of positive feedback." And Lovelock has the chapter and verse on this:

> Extra heat from any source, whether from greenhouse gases, the disappearance of Arctic ice or the Amazon forest, is amplified, and its effects are more than additive. It is almost as if we had lit a fire to keep warm, and failed to notice, as we piled on fuel, that the fire was out of control and the furniture had ignited. When that happens, little time is left to put out the fire before it consumes the house. Global warming, like a fire, is accelerating and almost no time is left to act.

Gaia, by the way, is the Greek name for Mother Earth, and Lovelock's idea here is that Earth is a living, conscious being that constantly adapts and adjusts itself to circumstances. Earth is not static, but alive and dynamic. Why, then, one might ask, can Earth not cope with the 0.01 percent of $CO_2$ being put into the atmosphere by human industries, as opposed to what nature itself puts there via causes such as the outgassing of carbon from the seas, where almost all the planet's carbon is stored? Humans are said to produce some 7.2 gigatons (Gt) of $CO_2$ a year, but the ocean has about 39,000 Gt of $CO_2$ dissolved in it, some of which becomes $CO_2$ in the air and some of which becomes limestone, joining the other 70 million Gt of carbon in Earth's rocks.

Yet even if Lovelock knows his earth sciences, as far as $CO_2$ goes, he clearly has his eye less on the wonderful complexity of nature and more on some 7 billion guilty individuals whom he seems rather sure must be punished.

In an interview with the British magazine the *New Scientist* in January 2009, Lovelock predicted that up to one-half of the world's population could die from runaway global warming in the period 2050 to 2100. By comparison, when Earth had just a billion people on it, Lovelock says, "their impact was small enough for it not to matter what energy source they used." This new twist on Malthusian doctrine (in Lovelock's version there are too many people and too few nuclear power plants!) cheerfully ignores, of course, all the contrary facts. Lovelock makes almost no reference to the fact that less than one-sixth of the world's population, living in the rich, developed economies, consume about one-half of the world's fossil fuels, while the vast majority of the world's population use the rest and consequently emit much less $CO_2$ per person.

A quick look at Greek myths, for example, shows the consequences of ancient peoples chopping down and burning forests on the face of Gaia. The planet was scarcely able to brush aside the first billion people, but eventually it did adapt. And so, you might think, it will continue to do so. You might think that, but Lovelock and the Green lobby do not.

Lovelock again: "As individual animals we are not so special, and in some ways are like a planetary disease, but through civilisation we redeem ourselves and become a precious asset for the Earth; not least because through our eyes the Earth has seen herself in all her glory."

For environmentalists like Lovelock, it is very satisfactory that climate change theory sees the folly of humankind as resulting from the Industrial Revolution. Environmentalists seem to hope that the sins of the past are—at last!—now going to have to be fully paid for.

Although there is much competition, Lovelock considers himself to be the voice of true environmentalism. Indeed, in a long article on the benefits of nuclear energy in *Reader's Digest,* he opines that "it was an invention of mine that kick-started the environmental movement." This was a gadget called the Electron Capture Detector, which measures air cleanliness, that he coinvented with another garage-based handyman. With it, Lovelock says, evidence of the spread of the pesticide DDT worldwide and of "chemicals called CFCs (chlorofluorocarbons) that were accumulating and damaging ozone in the atmosphere" emerged.

As it happened, the US government was unusually interested in this matter, perhaps because US companies held all the key patents to replacements for these CFCs (up till then used in household refrigerators and aerosol cans). This led to Lovelock working with the National Aeronautics and Space Administration, where he studied the question "Is there life on Mars?"—and that led him to climate change theory.

CFCs are now banned, and refrigerator companies in the United States are doing very nicely. Yet still "Mother Earth is in trouble." Lovelock writes again on behalf of suffering Gaia and, it seems, of many environmentalists.

> Every time we click a light switch or start a car, something sinister happens. From power station chimneys and car tail-pipes, immense volumes of gases such as carbon dioxide are pumped into the sky where they pollute the environment and act like a greenhouse, overheating the globe.

Fortunately, there is a much better option for Planet Earth; a new "green" solution. Lovelock again:

> A lifeline does exist and it's dangling in front of us. By grasping it now we can rescue the world from both the consequences of global warming and our looming energy shortage. It's safe, proven, practical and cheap. Our lifeline is nuclear energy.

To back this up, Lovelock briefly summarizes the problems with fossil fuels. To power a modern city, oil needs "a 1000 km [620-mile] line of railway trucks filled with expensive coal," which when burned will leave behind "500,000 tons of toxic ash." Oil "emits nearly as much greenhouse gas as coal plus huge volumes of sulphur and other deadly compounds that turn into acid rain." Gas (favored by the likes of Monbiot) is slightly better, but the supply is "vulnerable to terrorists." But finally on to nuclear. Here is a power source that "feeds on about two truck loads of

### (NOT) WHY VENUS IS UNINHABITABLE NOW

The greenhouse effect story starts with a Swedish (Nobel Prize–winning, actually) chemist named Svante Arrhenius, who at the start of the twentieth century observed through his telescope that Venus was totally obscured by clouds. He wrote in a book called *The Destinies of the Stars* that a "very great part of the surface of Venus is no doubt covered with swamps" with humid conditions not unlike the tropical rain forests of the Congo. With Arrhenius's bold sweep of the pen, Venus thus became, for much of the twentieth century, a place for science fiction films and writers to place all manner of unusual life-forms, from galactic dinosaurs to superintelligent carnivorous plants. Omniscient scientists often compared the planet to Earth in the Carboniferous Period. But years later, better technology began to reveal a rather less hospitable planet. Observations using spectrometers revealed an atmosphere consisting not of water vapor but almost entirely of $CO_2$. And the planet was much hotter than previously thought. *Hundreds* of degrees centigrade hotter. Too hot even for dinosaurs.

Shame! Thus it was that the June 1982 issue of *Popular Science* magazine proffered dire warnings of the effect of pumping too much $CO_2$ into the atmosphere, explaining: "Venus once had as much water as Earth. It lost the equivalent of Earth's oceans in the process of becoming a runaway greenhouse."

So runaway, indeed, that a block of lead placed on the surface would turn into a puddle. No wonder that Venus's beautiful seas boiled away long ago. Could a similar thing happen here? scientists were asked. The reply: The scenario is complex but seems to fit observations. Asked about Earth's own future, one farsighted chap warned that "the amount of carbon dioxide we're putting into Earth's atmosphere today is the most dangerous of all human activities."

cheap and plentiful uranium imported from stable countries like Canada and Australia." And the toxic waste? "A few bucketfuls."

It is not actually even that toxic, Lovelock adds. "The radiation from a reactor is tiny: about as much as that from our own bodies." According to the UK's National Radiological Protection Board, for example (which Lovelock thinks is a very good judge, despite the fact that it is less a public watchdog than the public relations arm of the nuclear industry), radiation doses from nuclear power stations amount to less than 1 percent of normal annual exposure from things like background radiation in rocks. "Compared with known cancer risk such as smoking and poor diet, the risk from non-medical, manmade radiation is about 1/100th of one percent."

### WHOEVER PAYS THE ORGAN GRINDER CHOOSES THE TUNE

In her book *Climate Money* (published in 2009 by the US Science and Public Policy Institute), Joanne Nova gives one of the first assessments of what $CO_2$ trading and carbon finance had cost the US government and therefore taxpayers over the years, using government-sourced documents for her research. Nova states that she discovered a "well funded [and] highly organized climate monopoly" based on highly selective or, as she puts it, "unaudited," scientific views, opinions, and theories.

Perhaps Nova's most scathing accusation is that the $79 billion that she calculates has been spent by the US governments on climate-related activities over the last 20 years has "created a powerful alliance of self-serving vested interests." In the best spirit of "Baptist and Bootlegger alliances," these interests are, in her opinion, compelled primarily by the lucrative profits to be garnered from emissions and carbon trading, when, or if, it becomes obligatory in the United States. As for the massive amounts of government cash poured into climate business since the end of the 1980s, even the $30 billion it spends on research into climate "science" has, to date, not produced "a single piece of empirical evidence that man-made carbon dioxide has a significant effect on global climate." What the research has produced, naturally, is a new breed of highly "environmentally concerned" scientists.

The economist Bruce Yandle coined the phrase "Baptist and Bootlegger coalitions" to describe cases where the economic interests of businesses and the moral concerns of campaigners coincide. The virtuous crusade to save the planet from runaway global warming, and the grubby business of selling nuclear power, are paradigmatic examples.

In this analogy, both Baptists and bootleggers want the sale of alcohol banned—but for different reasons. The Baptists want it banned because they consider alcohol to be morally wrong, while the bootleggers want it banned because then its price will rise and they can make easier profits. The Baptists would vehemently deny that they were assisting the bootleggers, just as Greenpeace and its partners in the Climate Action Network would bristle at the suggestion that they were assisting the nuclear industry and the oil companies with their holdings of natural gas, or even the expansionist instincts of states.

Bootleggers are, of course, completely amoral. Only money motivates them. Recall that all this nuclear-friendly advocacy was taking place in the context of the 1986 Chernobyl nuclear disaster, which spread a cloud of poisonous radiation over much of Europe and led to the ostensible "phasing out" of atomic power in many countries.

Equally, although skeptics point to ways that climate change policies may increase human suffering, not reduce it, what they do not realize is that some people *want* this increased suffering. For many environmentalists, like the most fiery Baptist preachers, humanity deserves to be punished for its poor stewardship of Earth. The aim is not to save people or even living creatures and forests but to save the *planet.* And that is a long-term business.

Climate change bootleggers do not limit themselves to promoting reactor sales; there are also huge profits to be made from wind farms and solar power (as long as the manna of government subsidies keeps flowing). Bootleggers proffer the same plausible sales spiel: Fossil fuels are dirty; renewables are clean. Coal is dirty and working class, conjuring images not only of gritty unionists going on strike but of working-class dads, as portrayed by D. H. Lawrence, returning home dirty, drunk, and in a filthy temper. Renewables, in contrast, and even nonrenewable nuclear power, are high-tech, clean, and modern, evocative of sunshine and fresh air. Overwhelmingly staffed by casual, nonunionized labor (contrary to the high-tech image), renewables are as close to economic nirvana as any surviving admirer of Margaret Thatcher or Ronald Reagan can get.

And in fact, all the most determined and most media-friendly environmental experts, whether these are advisors to the clique of hedge funds operated by Al Gore or the "environment skeptics" like Bjørn Lomborg or James Lovelock, are word perfect with their new enthusiasm for nuclear power. Green gurus recommend it because it is "safe, clean, and effective." Brushing aside all those nuclear accidents you may have heard about, they say that so-called civil nuclear energy from its start in the early 1950s has proved to be the safest of all energy sources. Lovelock puts it this way:

> We must stop fretting over the minute statistical risks of cancer from chemicals or radiation. Nearly one third of us will die of cancer anyway, mainly because we breathe air laden with that all pervasive carcinogen, oxygen. If we fail to concentrate our minds on the real danger, which is global warming, we may die even sooner, as did more than 20,000 unfortunates from overheating in Europe last summer [2003].

For such new nuclear romantics, disasters like Chernobyl are just media puffery. Lovelock even says that, in fact, "only 42 people died, and they were mostly firemen and plant workers." Firemen and plant workers! Such people are paid to get frazzled occasionally. They serve the cause of perfecting nuclear reactors.

The Green Guru goes on to add that since the explosion, UN experts have found no evidence of birth defects, cancers, or other health effects, "with one exception. Some 1,800 non-fatal thyroid cancers have been found in people who were children at the time. It is not even clear that they were triggered by the accident and they could have been avoided had the authorities issued warnings to stay indoors for 24 hours and issued iodine tablets." Other UN reports put the figure rather higher—about 8,958 people higher. But that is statistics for you.

Compare the pristine cooling pond at the heart of a nuclear reactor, or even the quietly ticking banks of electronic monitors in the control rooms, with the hellish disorder of a dismal pit in China where chunks of trees are burned under an orange and black sky to produce charcoal. During the run-up to the 2009 Copenhagen conference, the latter image was much reproduced to accompany articles about the need to reduce $CO_2$ emissions. Yet, ironically, converting wood into charcoal is actually a renewable, green process (if rather smoky), and in some countries it attracts government subsidies.

Ironic, too, that "renewable energy" as a term also includes biofuels, waste incinerators, and even dams across the Amazon, all projects that are hugely suspect environmentally. The public image, of course, is all solar panels and wind turbines, yet even these come with their own environmental problems. Solar panels are made with some of the most hazardous chemicals known to industry, such as arsenic, gallium, and cadmium for their semiconductor materials, a fact that helps explain why manufacturing plants are so expensive in countries with tough environmental and health regulations and why the panels generally are made by workers in developing countries. Wind turbines come with access roads and electricity pylons attached, not to mention thousands of tons of cement to anchor them, and offer their own peculiar intrusion into the landscape. But more to the point, neither of these sources can make energy in anything near the volumes that conventional power stations can. They simply cannot provide energy on the scale required to replace fossil fuels. The bottom line is that 99 percent of energy in the United States (for example) is provided by sources other than wind and solar, and that is not likely to change anytime soon. (International Energy Agency statistics give their combined role worldwide as about 1.25 percent of world energy.)

This is pretty complicated, so neutral observers can understand why so few journalists, let alone experts, can be bothered to report or discuss any of it. But let us at least agree on the polar bears.

When Al Gore made his critically acclaimed film, *An Inconvenient Truth* (2006), starring some polar bears stranded on melting ice off the coast of Alaska to support his claims about global warming, he used photos that had been taken by a marine biology student, Amanda Byrd, while she was on a university-related research cruise in August 2004, a time of year when the fringe of the Arctic ice cap *naturally* melts. It was later distributed by Environment Canada, a Canadian government department, to media agencies.

With an enlarged version of the polar bear picture on the screen behind him, Gore states, "Their habitat is melting . . . beautiful animals, literally being forced off the planet. They're in trouble, got nowhere else to go."

However, according to Byrd, when she took the picture, the bears did not appear to be in any danger, despite what its widespread use in worldwide media to illustrate coverage of global warming implied. An Environment Canada spokesman, Denis Simard, told the *National Post,* a national Canadian newspaper, that you "have to keep in mind that the bears aren't in danger at all. It was, if you will, their playground for 15 minutes. . . . [T]hey were not that far from the coast, and it was possible for them to swim."

The polar bear still is the symbol of the effects of global warming—but it is a cleverly designed marketing symbol, not a rational, scientific marker.

The pure white of both polar bear fur and the animals' dwindling iceberg home are contrasted with the nasty, dirty charcoal pits and dark, sinister black coal . . . the traditional "black/white" dichotomy. But in fact there is only one thing even purer than driven snow, and that is the gleaming, glowing heart of the atomic reactor. No wonder so many modern environmentalists have fallen in love with it.

## TOXIC SOLUTIONS?

If you still have a soft spot for "renewable" (a misleading term that covers ecologically disastrous hydroelectric schemes too), do not be too misty-eyed. Because wind turbines operate at such incredibly low efficiencies, and most of the power they do generate is at the wrong time and in the wrong place, before wind power could contribute 20 percent of, say, Britain's electricity, there would have to be about 100,000 wind turbines. That would require a dedicated six-mile strip right around the United Kingdom to make room for them. Appropriately, that would even displace Britain's one Green member of Parliament—in the resort town of Brighton.

Would solar energy be any better? No. The Sun is good at providing power for things like light-emitting diodes (LEDs) that light up at night along the side of a suburban driveway, but it is woefully inadequate for mass-power generation. Even to power a 7-kilowatt (kW) household heat pump would require about 1,227 square feet of standard photovoltaic arrays; to replace a small power station would require about six square miles of land. There are more efficient solar systems, but they use chemicals like arsine and phosphine. Arsine is almost as toxic as methyl isocyanate (which, when 40 tons were unintentionally released in Bhopal, killed 3,000 people and injured 200,000 more). The fact is, all solar photovoltaic systems involve toxic chemicals.

Speaking of deadly chemicals, fittingly, it was a chemist who helped sound the alarm about $CO_2$. Margaret Thatcher called in the scientists to help her come up with reasons to shut down Britain's coal industry. Not because she hated northerners—she came from Grantham herself, not so far from the Nottingham coalfields—but because British coal costs billions of pounds each year more to mine than foreign coal costs to import. She realized that if Greens could come up with spurious reasons to declare coal a deadly form of pollution, she could save the country money while appearing virtuous. Well, it did not happen quite like that, but the legacy is still there. Today Britain and Europe have dismantled their coal industries, burned most of their reserves of natural gas in privatized power stations, and now are about to run up hitherto unheard-of sums for windmills and tidal technology. And Europe is still heavily dependent on coal, but now it is imported—as much as 300 million tons a year—or about three-fifths of a ton for every single inhabitant of the Union.

# MYTH 3

# NUCLEAR REACTORS ARE RELIABLE AND SAFE

*Facts don't dominate, fear does.*

—John Hutsons, chief executive officer of Fresno Nuclear,

a $5 billion nuclear project in the United States

*Windscale Fire Aftermath—Seascale, Cumbria. "Scientists wearing protective clothing examine the temperature recorders on top of the reactor in the number one pile at Windscale, after the recent mishap," went the caption for this 1957 Press Agency photo, explaining that these rather dull-looking dials were the temperature recorders where the rise in temperature in "the pile" was first detected. (Press Association)*

HOW MANY ACCIDENTS DO THERE HAVE TO BE BEfore. . . ? The list is already long. It is accepted that nuclear power is potentially very dangerous, but we are regularly assured that there are so many safety systems with so many layers of security that the risk of an accident is a million to one. At the same time, the roll call of accidents and near misses gets longer and longer.

Here are just a few of the most famous accidents. There are plenty of others that no one wants to talk about.

**December 12, 1952.** The accidental removal of four control rods at an experimental nuclear power reactor at Chalk River, Canada, near Ottawa, led to a partial meltdown of the reactor's uranium fuel core. This is the first known major malfunction of a nuclear reactor.

**October 7, 1957.** When workers discovered a fire in Pile No. 1 at Windscale, the plutonium-production plant north of Liverpool in the United Kingdom, they sprayed it with carbon dioxide, a technique that was disastrously ineffective.* By the time the fire was put out with water, radioactive material had contaminated some 200 square miles of the surrounding countryside. "At the time, the Atomic Energy Authority rushed to assure an anxious public that radiation levels were still only a tenth of that considered dangerous. It was not until the 1980s that a report was released suggesting that the level of radiation near Windscale after the accident was up to 40 times greater than had originally been claimed." Many years later, the government

* The fire simply stripped the oxygen from the carbon.

grudgingly accepted that "at least 33" cancer deaths could be attributed to the effects of the accident.

**January 3, 1961.** A minor problem during the removal of the control rods at the core of a military-experimental reactor operated (like Windscale) to produce weapons-grade plutonium, near Idaho Falls, Idaho, quickly led to an explosive buildup of steam. Three servicemen were killed, one of them grotesquely impaled on a control rod. These deaths were the first publicly admitted "direct" fatalities in America from nuclear reactor operations.

**October 5, 1966.** The first and *only* attempt at a large-scale "commercial prototype" fast breeder reactor, the Enrico Fermi Atomic Power Plant, Unit 1 (Fermi 1) near Detroit, Michigan, came close to disaster after a zirconium plate at the bottom of a reactor vessel came loose during a test for full power, blocking the flow of liquid-sodium coolant and causing two fuel subassemblies to melt. Only four minutes passed from the first alarm to this meltdown. Fearing the release of radioactive iodine-131 from the secondary containment system, the operators considered evacuating nearby Detroit. Fermi 1 is particularly interesting, as it was designed to produce, or "breed," fuel by creating more fissile material than it consumed and thus achieve the "plutonium economy" that many people then, and now, see as the future of nuclear power. Fermi 1 was restarted in 1970 but had to be shut down again in 1972 when its core overheated again. (According to the US Nuclear Regulatory Commission, the fuel and blanket subassemblies were shipped off-site in 1973. The nonradioactive, secondary sodium system was drained, and the radioactive, primary sodium coolant was stored in tanks and drums until removal from the site in 1984.)

**March 22, 1975.** A worker using a lighted candle to check for air leaks at Browns Ferry reactor near Decatur, Alabama, touched off a fire that damaged electrical cables connected to safety systems and allowed the reactor's cooling water to drop to dangerous levels. Public statements by US nuclear authorities maintained that no radioactive material escaped into the atmosphere as a result.

**March 28, 1979.** Potentially the most serious US nuclear mishap took place at Three Mile Island. Loss of coolant caused radioactive fuel to overheat, leading to a partial meltdown and the release of a still-disputed quantity of radioactive material from the second reactor at the site. Partial decommissioning of this reactor is estimated to have cost at least $805 million.

**March 8, 1981.** A problem-ridden nuclear power station in Tsuruga, Japan, was found to have been leaking radioactive wastewater for several hours. Workers dispatched to mop it up were heavily irradiated. The problem was publicly disclosed only six weeks afterward, when radioactivity was detected in a nearby bay.

**January 4, 1986.** An improperly heated, overfilled container of nuclear material at the Kerr-McGee Corp. uranium-processing plant in Gore, Oklahoma, burst, killing one worker and allowing some radiation to leak out of the plant. More than 100 local residents required hospital treatment.

**April 1986.** An unexpected problem arose during routine testing of Reactor No. 4 at the Chernobyl site in the Ukraine . . . of which much more will be discussed later.

**July 2007.** An earthquake in Japan caused the Kashiwazaki-Kariwa plant to partially explode, releasing a cloud of toxic radiation that drifted, fortunately, out to sea. The accident tweaked aside the curtain of secrecy around Japanese nuclear power, and revealed another 97 incidents in the previous 30 years, 19 of which were considered by safety regulators to be "critical."

**March 2011.** The double blow of an earthquake followed by a tsunami left the Fukushima nuclear plant (again in Japan) without power and unable to prevent the partial meltdown of fuel rods at several reactors. Tens of thousands of people had to be evacuated from the surrounding area, and the United States publicly disputed claims by the Japanese operators that "everything is under control."

Actually, over the years, there have been many nuclear accidents at plants in Japan, most badly mishandled and clumsily covered up, which has only added to the public dislike and suspicion of the industry. Fukushima was no exception. Indeed, according to Arnold Gundersen, a former nuclear industry senior vice president, Japan now has the dubious distinction of being father to "the biggest industrial catastrophe in the history of mankind." Gundersen, a licensed reactor operator with almost 40 years of nuclear power engineering experience who manages and coordinates projects at 70 nuclear power plants around the United States, estimated that, as of summer 2011, the Fukushima nuclear plant likely had even more exposed reactor cores than was being admitted. "Fukushima has three nuclear reactors exposed and four fuel cores exposed. You probably have the equivalent of 20 nuclear reactor cores because of the fuel cores, and they are all in desperate need of being cooled, and there is no means to cool them effectively." Gundersen's assessment of the chances of solving the crisis is not encouraging either:

> Units one through three have nuclear waste on the floor, [derived from] the melted core, that has plutonium in it, and that has to be removed from the environment for hundreds of thousands of years. Somehow, robotically, they will have to go in there and manage to put it in a container and store it for infinity, and that technology doesn't exist. Nobody knows how to pick up the molten core from the floor, there is no solution available now for picking that up from the floor.

Today, nuclear scientists themselves joke about how, in the early days, they used to treat radioactive hazards and cut corners on safety, but the joke is on us. Secrecy is the best friend of the nuclear safety record, but the veil is tweaked aside when there is an "incident," whenever the worldwide system of radiation monitors (including those set up to check for secret atom bomb tests) goes off. Three days after the nuclear meltdown at Chernobyl in 1986, even as hundreds of thousands of Soviet citizens received potentially deadly doses of radiation, there had been no public announcement of the problem by the official Soviet news agency, TASS. Instead, the alarm was sounded 1,000 miles away by the government of Finland. Years later, Mikhail Gorbachev, the Soviet leader at the time, described Chernobyl as a turning point.*

Afterward, it was no longer possible for the Soviets to carry on in the same way of concealing corruption and incompetence under the heading of the national interest. *Glasnost* (or "openness") was not so much a controversial political strategy as an inevitable correction.

Today, Russian authorities have retreated from the philosophy of glasnost and, not coincidentally, rehabilitated their nuclear industry. With considerable panache (as mentioned in Myth 2), the Russian reactor company Rosneft even uses the 1986 disaster as a selling point, saying that the technical lessons learned in Chernobyl make its new reactors better than anyone else's!

A bit cynical, you might say. But for true cynicism, why not pop over to Japan, where, over the decades, the authorities have devoted a great deal of effort to reassuring the public about the safety of nuclear power. Part of this strategy has been converting visitor centers at nuclear complexes into quasi–holiday theme parks, with interactive games and cartoon characters. As journalist Norimitsu Onishi recently recounted, some of the juxtapositions are incongruous, to say the least. "It's terrible, just terrible," says a White Rabbit straight out of the children's book *Alice's Adventures in Wonderland.* "We're running out of energy, Alice!" An industrial robot then appears and reassures everyone that there is a new, safe, and very clean source of power, even a renewable one, if you build

---

* But as events were unfolding, Gorbachev denied there was a problem with the best of them.

reactors that recycle fuel. "Wow, you can do that?!" asks this inauthentic Alice, and then advises her rabbit friends that "You could say it's optimal for resource-poor Japan!"

But let us move back to the real world. In 2000, the Japanese Nuclear Industrial Safety Agency admitted to a scandalized public that, for over 25 years, it had falsified documents relating to safety inspections. It acknowledged 200 such falsifications, concealing 19 "critical" incidents and hiding 3 actual full-blown accidents, including one at the Fukushima complex. Following this mea culpa, the country's major nuclear operator, TEPCO, shut down no fewer than 7 of its 17 reactors, pending a new inspection and correction of safety defects.

After an earthquake in Japan in 2007 caused the Kashiwazaki-Kariwa plant to partially explode, the International Atomic Energy Agency (IAEA) criticized the lack of security at the site. In 2008, during a meeting of the Group of Seven industrialized countries, the body went further and warned that *all* Japanese reactors were unfit to cope with major earthquakes. After all, Japan is famously on the Pacific "Ring of Fire"—that is, a fault line between two major tectonic plates. Nonetheless, the Japanese regulator sets for its plants a very low bar on safety—they need to withstand only earthquakes up to 7.0 on the Richter scale. (The earthquake that devastated Fukushima was about level 9.0 by comparison.) As is customary, the IAEA's warning, however, was issued privately, and came to light only as part of the release of confidential documents by the Internet site WikiLeaks in 2010.

Although the IAEA is famous for its power to stop states from building nuclear bombs, it has no power to stop them from building dirty bombs—in the form of dangerous nuclear reactors that "do a Chernobyl." Its Operational Safety Research Team has no authority over national regulators and must content itself with timid recommendations given in private—friendly advice that the operators can choose to ignore, and often do.

Take the case of the new so-called evolutionary power reactor (EPR), to use the name it registered with the US safety authority, the one promoted as "the safest reactor ever." The French have applications in to

### REGULATORS NOT REGULATING, JUST WAVING

- **United States.** According to Frank N. von Hippel, a nuclear physicist responsible for national security issues in the White House Office of Science and Technology Policy from 1993 to 1994, in the United States, a dangerous custom has developed whereby only supporters of the nuclear industry are allowed to supervise it. Lobbyists have been allowed to have an effective veto over regulators who have ever been publicly critical of the industry.
- **China.** In China, Kang Rixin, former general manager of the state-owned China National Nuclear Corporation, was sentenced to life in jail in 2010 for accepting bribes (and other abuses). The verdict raises questions about the quality of his work on the safety and trustworthiness of China's nuclear reactors.
- **India.** In India, the nuclear regulator reports to the national Atomic Energy Commission, which champions the building of nuclear power plants there. The chairman of the Atomic Energy Regulatory Board, S. S. Bajaj, was previously a senior executive at the Nuclear Power Corporation of India, the company he is now helping to regulate. Interviewed by the *New York Times* in 2011, Bajaj said, rather nicely, that his agency was "functionally independent."
- **Japan.** In Japan, the regulator reports to the Ministry of Economy, Trade and Industry, which overtly seeks to promote the nuclear industry. Ministry posts and top jobs in the nuclear business are passed among the same small circle of experts, a practice that the Japanese have a word for: *amakudari,* or "descent from heaven." As Eisaku Satō, the sometime governor of Fukushima province (with its infamous nuclear reactor complex), puts it: "They're all birds of a feather."

build these in Normandy, on the French coast, and in Finland, Britain, and the United States, applications that have been delayed by wrangling over . . . safety.

Specific questions concern overheating and fracturing of the control rods at the heart of the reactor and of the ability of essential machinery to withstand small leaks of radiation. Leaked documents reveal that engineers had found that the control rods failed to eject normally when temperatures in the reactor core rose and in a minor accident, nearly 33 percent of them would break, whereas safety regulations set the limit as 10 percent.

Control rods, as the name implies, are crucial. Insertion and removal of these fuel rods regulate the rate of the nuclear reaction in the reactor vessel. A problem ejecting one rod has a domino effect, causing the reactor to overheat, which in turn can melt the thin metal cladding on other radioactive fuel rods, causing them to release radioactivity. When enough rods have problems, the entire core can go into meltdown.

Challenged over the problems, a spokesperson for one of the huge French corporations behind the design, Électricité de France S.A. (EdF), merely said that the documents were working papers and that, of course, "Nuclear safety is EdF's top priority." Yet so too is the need to keep the cost of the plant down.*

Applications to build EPR plants in Britain and the United States have been delayed by requests by national regulators for more investigations of these safety issues. But back home in France, the regulator approved the Normandy reactor without asking for more research. According to Guillaume Wack, director of the French safety directorate itself, "the cost of detailed testing would be too high for the industry to finance without the certainty of having at least one launch order."

In Finland, where Teollisuuden Voima Oyj (TVO), the Finnish utility, is actually building the first one of the new reactors, disputes have been more difficult to resolve. According to Areva-Siemens, the Franco-German consortium behind the reactor, TVO does not trust it to modify the fiendishly complex design as it sees fit; instead TVO demands documentation and approval from regulators for every change, however

---

* As mentioned, essential machinery in nuclear plants is supposed to be able to withstand damage to at least 10 percent of the rods. But one leaked document, dated April 2009, warned that some of the equipment used in the EP design would fail if just 1 percent of the rods were to break, leaving the plant potentially unable to shut down the reactor core in the event of even a small malfunction.

small. Indeed, in a warehouse beside the site is a very respectable library consisting of two and a half miles of shelving containing some 160,000 documents. (Nuclear reactors, like rocket travel, generate vast amounts of paperwork.) However, TVO complains that Areva is treating the new reactor as a research and development project in which the Finns themselves are guinea pigs.

Speaking of dangerous experiments, the fuel for the new reactors, called mixed oxide fuel (MOX), is also especially risky, as it is a mix of plutonium and uranium oxides. Cheaper than the uranium fuel used in most reactors, it is more highly irradiated and hence harder to use, store, and cope with in the event of an accident. According to Wack, the French regulator, the motivation to use MOX was purely economic. And of course he had no problem with that.

In Japan, the regulators and the regulated have long been friends, working together to offset the doubts of a public brought up on the horror of the nuclear bombs on Hiroshima and Nagasaki and the hundreds of thousands of radiation deaths that followed. Indeed, cancer rates are still higher in the prefectures where the bombs were dropped during World War II than elsewhere in Japan. Falsifying safety reports had been part of the cozy relationship between regulators and regulated. It stopped only in the mid-1990s after a whistleblower working for a separate company employed to inspect the reactors reported that TEPCO was in the habit of hiding cracks in its reactors—it later emerged that there were even cracks in the steel shrouds that cover reactor cores.

What was the Japanese reaction to the whistleblower's report? The Japanese regulator simply passed the report on to the plant owner, along with the informant's name (he never worked for the company again!), and allowed the plant to continue to operate. Commenting on nuclear regulation in Japan after Fukushima, Eisaku Satō, the governor of the province at the time that the secret cracks were revealed, was appalled that the regulators had allowed TEPCO to continue operating. "An organization that is inherently untrustworthy is charged with ensuring the safety of Japan's nuclear plants . . . the whole system is flawed. That's frightening."

If that is not enough, in Japan, the electoral system allows parties to reward large institutional backers, such as the nuclear industry, with parliamentary seats. This is how Tokio Kano, a vice president of Tokyo Electric, was given two terms in the Upper House of the Japanese parliament. While there, he held key positions on energy committees and reshaped the country's energy policy by putting nuclear power at its center, citing the need to "reduce carbon emissions (as mandated by the Kyoto Climate Change targets)" and nuclear's fabled promise of "energy independence." He pushed through policies dear to the industry, such as the approval of MOX fuel in reactors, despite its particular dangers. He opposed attempts to open up the power industry to more economic competition. He even attempted to have textbooks rewritten to include positive statements about nuclear power; in his words, "Everything written about solar energy is positive, but only negative things are written about nuclear power."

Today, a similar media war is under way to reassure the Japanese—and indeed people in places as far away as California—that the radiation effects of the Fukushima incident are not dangerous. Little seems to have changed since the time of Chernobyl, when the Soviet authorities decided that the radiation hazard was limited to a small circle around the plant, affecting the relatively small number of 10,000 people—who were all then forcibly evacuated. Twenty years on, a multiagency panel of the United Nations calculated that, in fact, *200,000* square miles of eastern Europe were blanketed with invisible fallout, an area that included 5 million people. But at the time, the potential victims were given no guidance. It is from this population that most of Chernobyl's eventual victims came.

According to a 1996 study by the Moscow-based Center for Environmental Policy, "Chernobyl: Consequences of the Catastrophe for People and the Environment," levels of the deadly plutonium-238 isotope were 90 times higher in the region than they should be and traces of radioactive cesium-137 affected local fish and wildlife. But the clearest evidence of the hidden effects of nuclear fallout would eventually come in levels of thyroid cancer. The thyroid gland readily absorbs radioactive iodine, one of the elements of the Chernobyl fallout, and for years after the disaster, levels of this unusual cancer (unusual, that is, in terms of possible causes) were 500 times higher in the area nearest the plant and 52 times higher across Ukraine than the pre-1986 levels. In neighboring Belarus, where the radioactive cloud first blew, the study found that the levels were even worse—113 times higher.

To make matters worse, some scientists feel this risk could have been greatly reduced had the population known to take two simple precautions: suck iodine tablets and not drink local milk. But they never received such instructions. Instead, the authorities insisted that everything was under control. Similarly, today, chance discoveries of dangerous levels of radioactive cesium and other elements in Japanese beef, green tea, drinking water, and even on its famous cherry trees are quickly explained as "under control"—but then, of course, it would be hard to ban everything.

Recall that toxic cloud over Europe in 1986 that briefly terrified decision makers, leading them to destroy farm produce and pour not only milk but money down the drain. Here again a curious alliance of environmentalists and business interests has rewritten history. James Lovelock, the Green Guru, reassures us here again, for example, writing: "The fall-out from the radioactive cloud that swept Western Europe was really nothing: only one-tenth of a chest x-ray or ten days on holiday in the Alps."

So why on Earth (or on Gaia) do people worry about it? Why are we so frightened? asks Lovelock. Naturally, he blames the media and the "bad Greens" who spread irresponsibly negative opinions about our nuclear future. Expanding on this theme in a London daily newspaper, Lovelock added: "Opposition to nuclear energy is based on irrational fear fed by Hollywood-style fiction, the Green lobbies and the media." After all, he adds, if nuclear power was really so dangerous, then France, with 58 nuclear power stations and a power grid largely based on them, would likely be beset by pollution problems. "Far from it, the world's nuclear champion is safe and its health is among the world's best."

Lovelock is a veritable font of pronuclear facts, but let us stop him there. The thing about Chernobyl was not only how bad it was (which was awful) but how bad it *might* have been had it not been for the kamikaze heroics of those firemen and other 30,000 workers who participated in the damage-limitation work. The issue is not just what *did* happen but what so nearly *might* have happened.

The accident in Reactor No. 4 at the Chernobyl nuclear power station took place on the night of April 25, 1986, during a test. The operating crew planned to test whether, in the event of a loss of power, the slowing steam turbines could still produce sufficient energy to keep the coolant pumps running until the emergency diesel generator was activated.

In order to prevent an interruption of the test run of the reactor, workers deliberately switched off the safety systems. During the test, the reactor was shut down to just one-quarter of normal power. However, for unknown reasons, the power level fell to almost nothing. The plant staff therefore had to increase it again rapidly. There was a sudden and unexpected power surge, and within fractions of a second, both the power level and temperature rose many times over. The reactor, now out of control, exploded, blowing off the 1,000-ton sealing cap on the reactor building. As temperatures in the reactor core reached over 4,530°F, the fuel rods melted. The graphite covering of the reactor then ignited, and in the ensuing inferno, radioactive fission products were sucked up into the atmosphere, forming a deadly cloud that soon stretched for miles.

On April 27, only 36 hours after the accident, the 45,000 inhabitants of Pripyat, about four miles away from the reactor, were evacuated in buses. The town remains uninhabited to this day. In the period up to May 5, many more people living within a radius of 18 miles around the reactor had to leave their homes. Within ten days, 130,000 people from 76 settlements in the area were evacuated.

The authorities desperately tried to put out the fire at the blazing reactor building, trying first with water, which firefighters sprayed into the core of the reactor in the first few hours after the accident. Unfortunately, water is not suitable for fires in out-of-control nuclear reactors. The tremendous temperatures inside reactors immediately turn water to steam, which dissociates to hydrogen and oxygen and explodes. The next attempt was a more sophisticated effort by some 30 or so military

### WHOSE FAULT WAS IT ANYWAY?

The design of the reactor at Chernobyl utilized technologies that made rapid shutdowns particularly risky. The key to speeding up or slowing down the nuclear reaction in all reactors is the progressive insertion or removal of control rods made of a material that absorbs neutrons and thus slows the atomic reaction in the surrounding fuel. However, graphite moderators and control rods in the Chernobyl reactor had the effect of *increasing,* rather than dampening down, the neutron exchange in the first few seconds after the rods were inserted into the core and thus sharply raising reactor power output rather than reducing it, as desired. Apparently the reactor operators did not know about this counterintuitive behavior. Thus, in the emergency, their attempts to deal with the sudden surge of temperatures in the core actually made the situation much worse. For this and other reasons, the first official report on the accident blamed the operators; even a second, more generous report pointed at the deficient "safety culture" of the time. Neither report offers any convincing explanation of *how,* even had they known of the control rod problem, the staff that night could have possibly attempted to shut down the reactor safely.

helicopters to drop several thousand tons of boron powder and lead pellets and over 1,800 tons of sand into the burning reactor building to try to smother the fire and absorb the radiation.

These efforts were equally unsuccessful. In fact, they also made the situation worse. Heat accumulated under the dumped materials, the temperature in the reactor rose again, more of the reactor building's concrete melted and collapsed, and the quantity of radiation escaping increased. In one last-ditch effort, workers built a tunnel under the reactor (with dozens of them losing their lives in the process) through which nitrogen gas was passed to cool the reactor core. By May 6, ten days after the disaster had begun, both the fire and the radioactive emissions were finally brought under control. But by then, 400 times more radiation had escaped than after the Hiroshima atomic bomb.

The 600 men of the plant's fire service and operating crew were the most severely irradiated group. Of them, 134 received doses of radiation up to 13,000 times higher than the maximum "safe" dose of radiation to which individuals living near a nuclear power station should be exposed.

Although Lovelock says that radiation is not too bad, nonetheless, 31 workers died shortly after Chernobyl exploded. More than half a million people, mainly miners and soldiers, were involved in the cleanup operations. (In neighboring Belarus, 70 entire villages were demolished and then buried under mounds of earth, for example.) Many of these workers suffered health damage; 300,000 are believed to have received doses of radiation of more than 500 times the assumed safe level. How many of them have died to date as a result is a controversial question, but according to government agencies in the three former Soviet states directly affected, about 25,000 of these "liquidators," as they were called, have died so far. (Other estimates triple the official figures.)

Japanese schoolbooks, like schoolbooks the world over, do not dwell too much on Chernobyl, which is maybe a shame in light of recent developments, as accounts like that of Russian journalist, Svetlana Alexievich, could give children an invaluable insight into the horror of the nuclear disaster. In her oral history *Voices from Chernobyl,* Alexievich includes interviews with people like Lyudmilla Ignatenko, wife of fireman Vasily Ignatenko.

> *We were newlyweds. We still walked around holding hands, even if we were just going to the store. I would say to him, "I love you." But I didn't know then how much. I had no idea. We lived in the dormitory of the fire station where he worked. There were three other young couples; we all shared a kitchen. On the ground floor they kept the trucks, the red fire trucks. That was his job.*
>
> *One night I heard a noise. I looked out the window. He saw me. "Close the window and go back to sleep. There's a fire at the reactor. I'll be back soon."*

Mrs. Ignatenko did not see the explosion itself, of course, but she remembers the sight of the flames from the burning reactor.

*Everything was radiant. The whole sky. A tall flame. And smoke. The heat was awful. And he's still not back. The smoke was from the burning bitumen, which had covered the roof. He said later it was like walking on tar. They tried to beat down the flames. They kicked at the burning graphite with their feet. . . . They weren't wearing their canvas gear. They went off just as they were, in their shirt sleeves. No one told them.*

*At seven in the morning I was told he was in the hospital. I ran there but the police had already encircled it, and they weren't letting anyone through, only ambulances. The policemen shouted: "The ambulances are radioactive, stay away!"*

In the hospital, she could scarcely recognize her husband, so puffed up and swollen was he. She could barely see his eyes. A friend who worked there told her that he needed milk, lots of milk. They should drink at least three liters each. "But he doesn't like milk," Lyudmilla exclaimed. "Well, he will drink it now," her friend replied. Alexievich reports that many of the doctors and nurses in that hospital, and especially the orderlies, would later themselves get sick and die. But at the time no one knew what was in store for them. In any case, later that evening, Mrs. Ignatenko was not allowed entry into the hospital.

*The doctor came out and said, yes, they were flying to Moscow, but we needed to bring them their clothes. The clothes they'd worn at the station had been burned. The buses had stopped running already and we ran across the city. We came running back with their bags, but the plane was already gone. They tricked us.*

She was briefly allowed to see her husband later on. Only now he looked "funny," she thought, in pajamas several sizes too small with sleeves too short and his feet sticking out. At least his face was no longer swollen. It seems the patients had all been given some sort of fluid. It looked like the first sign of progress. She joked: "Where'd you run off to?" and her husband moved to hug her. But the doctor would not let him. "Sit, sit," she said. "No hugging in here." Lyudmilla continues:

*On the very first day in the dormitory they measured me with a dosimeter. My clothes, bag, purse, shoes—they were all "hot." And they took that all away from me right there. Even my underwear. The only thing they left was my money.*

But for Fireman Ignatenko himself—and the others—there was no easy way to get rid of absorbed radiation.

*The burns started to come to the surface. In his mouth, on his tongue, his cheeks—at first there were little lesions, and then they grew. It came off in layers—as white film . . . the color of his face . . . his body . . . blue, red, grey-brown. And it's all so very mine!*

The thing that saved her, she says, was that it all happened so fast that there was not any time to think and "there wasn't any time to cry."

*It was a hospital for people with serious radiation poisoning. Fourteen days. In fourteen days a person dies.*

Seven months after the explosion, the ruined reactor building and its molten core were enclosed within a reinforced concrete casing. This sarcophagus was supposed to absorb the radiation and contain the remaining fuel. It was considered to be an interim measure, with a designed lifetime of only 20 to 30 years. (Well, it's a start! Only another 10,000 or so more to sort out later.) Hastily constructed, there was a particular risk of the iron beams within it rusting.

In 1997, the major industrialized economies, led by France and together with the European Bank for Reconstruction and Development, launched a "Shelter Implementation Plan." The new shelter was intended to safely confine radioactive substances for at least 100 years. The 20,000-ton structure was supposed to be completed in 2008 at a cost of $1 billion. As with most things nuclear, however, costs have spiraled upward while completion dates have been pushed back. (The shelter is now set to be completed in 2013.)

## CHERNOBYL'S LEGACY

In the first few years following the Chernobyl disaster, public opinion in Belarus and Ukraine was highly critical of any further expansion of nuclear power. A moratorium, declared in 1990, put further construction and extension plans on any new nuclear plans temporarily on hold. This moratorium was lifted two years later, however, driven by rising oil prices and the collapse of the Soviet Union, which stopped the supply of "friendship oil" at friendship prices.

And Japan's nuclear utilities will no doubt take heart from the fact that Ukraine's official energy policy today, just as it did before the disaster in 1986, counts the country's 15 reactor units as the key pillars of the energy supply, supplemented by natural gas and coal. Ukrainian reactors are still providing a similar proportion (40 percent) of the country's electricity as Japan's reactors did before Fukushima. "We do not have any significant reserves of other fuels and cannot therefore do without nuclear power," explained Ivan Plyushch, the then chairman of the parliament of Ukraine, at a memorial ceremony for the victims of the catastrophe on April 26, 2002.

In the former Soviet Union, nuclear "incidents," of which there were many, were always military secrets, but in resolutely demilitarized Japan, there is a thornier, cultural problem with accidents: Admissions of error are shameful. In either country, debates over safety were kept well out of the eye of the Japanese public, with their long-standing dislike of, indeed phobia about, all things nuclear.

The truth is, the Japanese have always had a particular reason to be fearful of nuclear power. The country sits on the edge of several earthquake faults known as the Pacific Ring of Fire. The last major earthquake there prior to the 2011 one, in 2007, off the shore of Niigata Province, also damaged a nuclear plant.

A key existential problem for nuclear reactors—besides the problem that they produce waste that is deadly for thousands of years—is that they are very hard to turn off. Even after the nuclear chain reaction has been stopped and the reactor is shut down, the fuel continues to burn, producing about 6 percent as much heat as it would when running. This heat is caused by continuing radioactivity producing subatomic particles and gamma rays, which in turn set off atomic reactions.

Usually, when a reactor is shut down, cool water from a river or the ocean is used to remove excess heat from the reactor core. Immediately after the March 2011 earthquake, the Japanese reactor complex shut down smoothly enough, but the atomic fuel, as ever, continued to burn and produce heat. Yet, with the reactor shut down, internal systems for cooling the reactor were out of commission. So the Japanese tried everything else they could think of: They sent police trucks mounted with water cannons to spray water into the reactor buildings. They had military helicopters drop water onto the roof of the complex—a "performance, a kind of circus" aimed more at reassuring the public than actually doing anything practical, as Ken-ichi Matsumoto, an aide to Prime Minister Naoto Kan, put it later. Finally, they borrowed a huge industrial pump from China (the indignity!) and tried to pump in seawater from outside—but because the area containing the fuel (known as the containment vessel) was already full of high-pressure steam, this was rather "like trying to pour water into an inflated balloon," as one expert put it.

All the while, the operator of the plant, the Tokyo Electric Company, reassured the Japanese public that "everything was under control." Yet this was hardly the usual procedure. As the world's press reported, in the desperate last hours of the stricken plant, the last few remaining staff would

> crawl through labyrinths of equipment in utter darkness pierced only by their flashlights, listening for periodic explosions as hydrogen gas escaping from crippled reactors ignites on contact with air.
>
> They breathe through uncomfortable respirators or carry heavy oxygen tanks on their backs. They wear white full-body jumpsuits . . . that provide scant protection from the invisible radiation. They have volunteered, or been assigned, to pump seawater on dangerously exposed nuclear fuel . . . to prevent meltdowns that could throw thousands of tons of radioactive dust high into the air.

# MYTH 4

# NUCLEAR ENERGY IS "TOO CHEAP TO METER"

*Perhaps the most, indeed only, really successful form of atomic power is also the smallest. In this archive news photo, a tiny atomic battery, whose basic material is a radioactive waste by-product called promethium-147, is powering a radio transmitter. The original caption described the nuclear battery as producing energy for "almost limitless periods of time" as well as requiring virtually no shielding against radiation; indeed, the battery gives off "less radiation than radium dials on regular wrist watches." The technology can adapt to almost all transistor circuits and was expected to be especially useful in . . . guided missiles.*

*Limitless power arrives, courtesy of the nuclear battery. (Press Association)*

NUCLEAR POWER HAS NEVER BEEN ECONOMICAL. The real question is: Just how much more does it cost? The answer is surprisingly difficult to pin down.

In the early days, nuclear reactors were financed by the military, since their purpose was to make fearsome weapons. In more recent years, the nuclear industry has been less transparently funded by public subsidies, tax waivers, cheap government loans, and other guarantees. The public is expected to pick up the cleanup costs, both when plants melt down (as at Chernobyl and Fukushima) and to keep nuclear waste in what are basically eternal repositories. Yet even with all these subsidies, nuclear power is increasingly crippled by its costs, even in countries with the most intensely nuclear-friendly governments, such as France.

The paradox of nuclear power is that the basic costs of running nuclear reactors are relatively low, yet the real cost of electricity produced—including repayment of construction costs—is substantially higher than traditional systems. As energy analyst Steve Thomas puts it:

> [O]nce a nuclear power plant has been built, it may make economic sense to keep the plant in service even if the overall cost of generation, including the construction cost, is higher than the alternatives. The cost of building the plant is a "sunk" cost that cannot be recovered and the marginal cost of generating an additional kilowatt-hour (kWh) could be small. Differences in assumptions on, for example, operating performance and running costs, which are not readily apparent in the headline figures, create competing realities and go a long way to explain how the two sides in the nuclear debate find so little common ground.

The existential economic problem for nuclear power is that the *real* price of electricity has been falling for decades while its generating costs continue to rise each year.

Think of it as buying a house while relying on bank loans. The regular monthly payment to the bank for the loan is the one that makes or breaks the family budget. If, one month, the family finds itself short of money, the parents can turn down the thermostat and tell the kids to keep off the phone, but they cannot ask the bank to wait for its check. It is the same with nuclear power companies. The fixed costs—that is, the costs that will be incurred whether the plant is operated or not—must be covered. And it is these fixed costs that essentially determine the price at which nuclear electricity must be sold. So if the fixed costs are high, every month the plant must charge a lot for its electricity, lest it cannot pay back the banks the next month. *But there's a catch: If the price of its electricity is too high, no one will want to buy it.*

Of course, there are ways around that. One is to make sure consumers do not know how much they are paying for their nuclear electricity. And since, for most countries, nuclear electricity is just a drop in the ocean of total electricity demand, it used to be easy to offer nuclear power companies guaranteed sales at preferential rates. It is unfortunate (for the nuclear industry) that market forces apply to competing power companies too. This liberalization of electricity markets has been a disaster for the nuclear industry. In fact, the Reagan and Thatcher administrations, which promised a strong revival in the nuclear industry, presided over steep declines in the fortunes of nuclear power.

**JARGON BUSTER**

To allow comparisons between reactors with different output capacities, costs are often quoted as a cost per installed kilowatt (kW). Thus, a nuclear power plant costing $2 billion with an output rating of 1,000 megawatts (MW) (i.e., 1,000 MW, which is equal also to *one million* kilowatts, would have a cost of $2,000 for each kW ($2,000 per kW).

The usual rule of thumb for nuclear power is that about two-thirds of the generation cost is accounted for by fixed costs, the main ones being the cost of paying interest on the loans and repaying the capital,

### NUCLEAR CASE STUDY: OLKILUOTO

The Olkiluoto nuclear power plant in Finland order is usually described as a turnkey contract, in which everything necessary has been included and the plant will be ready to go, all for a fixed, agreed price. The French industrial conglomerate Areva was responsible for management of the construction.

The €3 billion deal in 2003 was a huge boost for the nuclear industry in general and for Areva in particular. When the Finnish government signed up for the fancy new 1,600 MW pile, it was the first nuclear order in western Europe and North America in ten years (since the Civaux-2 order in France) and the first order ever (outside the Pacific Rim) for one of the new, more sophisticated (and hence expensive) generation III reactors.

The deal was seen as doubly important for the nuclear industry because it seemed to contradict conventional wisdom that energy market liberalization and nuclear power orders were incompatible. In fact, the Finnish electricity industry had been attempting to obtain parliamentary approval for a fifth nuclear reactor since 1992. But, sure enough, the dream soon turned sour. By March 2009, the project was acknowledged to be at least three years late and €1.7 billion over budget. In August 2009, Areva acknowledged that the estimated cost had reached €5.3 billion, which represented a cost of $4,500 per kW, and an acrimonious dispute between Areva and the customer, Teollisuuden Voima Oy (TVO), was raging.

but an estimate of the cost of decommissioning the plant is generally also included. The main bills in the operation of a nuclear plant are for things like staffing, the energy and materials needed to run it, and for maintenance and repair rather than for the uranium fuel.

Areva, for example, offers as a general rule of thumb that 70 percent of the cost of a kWh of nuclear electricity is accounted for by the fixed costs from the construction process, 20 percent by fixed operating costs, and a mere 10 percent by "variable" operating costs, such as the uranium fuel. (These statistics, of course, ignore end-of-life costs.) When you read that the price of nuclear electricity is such-and-such, whether it is very high or very low, the figure is likely meaningless. The "price" of electricity is as vague as the value of a spare room that could be rented out might be for the householder.

At least with houses, there is a price tag attached, whether they are bought new or bought from someone else. But with nuclear reactors, quoted construction costs, both forecasts and "actual," should be treated with skepticism. If you build yourself a new house, the construction firm provides you with a very precise bill. International nuclear power companies usually are not required to publish properly audited construction costs, and they certainly have little incentive to present their performance in anything other than a good light. In the United States, utilities are required to publish reliable accounts of nuclear plant construction costs for the economic regulator (as cost recovery is allowed from consumers only for properly audited costs). Not so in Europe. However, a glimpse into the economic realities of nuclear power is provided by the United Kingdom's Sizewell B plant, which was constructed by a company that had almost no other activities in which to disguise the costs. Steve Thomas again:

> It is obvious that prices quoted by those with a vested interest in the technology, such as promotional bodies, plant vendors (when not tied to a specific order), and utilities committed to nuclear power are as reliable as small ads for lovers in the magazines. Less well appreciated is that the prices quoted by international agencies, such as the Nuclear Energy Agency, must also be treated with care, particularly as they are usually based on indicative rather than real costs. These are figures provided by national governments, generally with political imperatives to show nuclear power in a good light, based not on actual experience, but on official estimates and "projections."

The much-discussed nuclear renaissance, a popular narrative at least up until the spring of 2011, is much overstated. The insurmountable problem for the industry is not environmental or even health-related (supporters never tire of citing supposed death tolls from inhaling sulfur particles from coal power stations) but economic. Ironically, it was here that the early nuclear supporters staked their claims. Over the years, in order to continue doing so, their manipulations of the figures have become increasingly extreme.

## 12 BEST TRICKS OF NUCLEAR ECONOMICS

### Trick 1: Make a Low Bid

The construction cost trick plays a major role in any estimate of the cost of power from a nuclear plant. Conventionally, such costs include the cost of the first charge of fuel but do not include the interest incurred on borrowings during the construction of the plant (usually known as interest during construction).

Construction costs forecasts are notoriously inaccurate. It is one reason why the World Bank has a standing policy not to lend money for nuclear projects. For example, according to a UK House of Commons Energy Select Committee report, the cost of the Sizewell B reactor was 35 percent higher in real terms than the price quoted when it was ordered in 1987. As a consequence, the cost of the electricity the plant produced went up to $5,400 per kW.

A number of factors make forecasting construction costs difficult. First, all nuclear power plants currently on offer require significant on-site engineering, which might account for about 60 percent of total construction cost. The major equipment items—such as turbine generators, steam generators, and reactor vessels—account for a relatively small proportion of total costs.

Partly in response to this, some nuclear power plant designs have been offered to possible buyers at a price that the vendor guarantees will not increase. And you do not need to be a bank manager to guess how this works out in practice. In the mid-1960s, the four major US nuclear vendors sold 12 plants under such terms. They got the contracts, then promptly lost massive amounts of money.

These days, the nuclear industry accepts that at *around $6 billion each, nuclear reactors cost at least four times more than a gas-fired plant.* So they tend to use arguments other than cost to sell them.

### Trick 2: Buy Five, Get One Free

The expectation with most technologies is that successive generations of design will be cheaper and better than their predecessors because of new learning, economies of scale, and technical changes. In the 1970s, major reactor vendors were receiving up to ten orders per year. This supposedly allowed them to set up efficient production lines to manufacture key components and establish skilled teams of designers and engineers.

Indeed, when the Performance and Innovation Unit of the Cabinet Office in the United Kingdom examined the economics of nuclear power in 2002, it used forecasts of costs from British Energy (the nuclear power plant owner) and British Nuclear Fuels Limited (the plant vendor) that were allegedly based on "a substantial learning and scale effects from a standardised program."

Yet a Nuclear Energy Agency (NEA) report from 2000 suggests that the intuitive expectation that economies of number would be large may be misleading in the case of the atomic power plants. It stated:

> The ordering of two units at the same time and with a construction interval of at least twelve months will result in a benefit of approximately 15% for the second unit. If the second unit is part of a twin unit the benefit for the second unit is approximately 20%. The ordering of additional units in the same series will not lead to significantly more cost savings. *The standardization effect for more than two units of identical design is expected to be negligibly low.* [Emphasis added.]

Trick 2, promising though it was, is now history because the major reactor vendors have received only a handful of orders in the past 20 years; their own production lines have closed, and they have cut back on the number of skilled teams they employ. Westinghouse has received only one order in the past 25 years, while Areva's order from Finland was its first in about 15 years.

## A FINANCIAL MELTDOWN OF ÉLECTRICITÉ DE FRANCE

Nuclear generation in France is approaching a crisis point. Of the 58 nuclear reactors in that country, 37 will be 30 years old by 2015, each requiring a three-month outage for inspection prior to licensing for their last ten years. For their operation to be extended for a further 10 to 20 years (before retirement at age 40) thereafter, they will have to be upgraded or replaced—perhaps with shiny newly built evolutionary power reactors (EPRs). And all this by 2020. Unfortunately, according to the nuclear energy analyst John Busby, the International Atomic Energy Authority PRIS (Power Reactor Information System) database for France shows that as of the summer of 2011, only two out of the 37 official French nuclear agency inspections due by 2015 had been completed.

Meanwhile, the French electricity utility Électricité de France (EdF) has accrued debts of over €50 billion, is unable to fund an upgrade or new-build program, and is prohibited by the French government from raising tariffs to fund its future. Only 1 of the 20 new-generation reactors essential to maintaining electricity is even under way (which, as we have seen, is a far cry from operating, anyway) at Flamanville. And Areva has yet to commission the first EPR in Finland. If it encounters similar problems in China, where it is supplying two EPRs and hopes to sell two more, Areva's debts will increase enormously.

Meanwhile, in a linked and potentially deadly spiral, the cost of borrowing money goes up. Gone are the old days when the debts of nuclear energy in France were fully backed by the French government with its AAA debt rating.

Instead, in June 2010 for example, rating agency Standard & Poor's cut Areva's debt rating to BBB+, the third-lowest investment grade, saying that the manufacturer's finances were strained by delays and cost overruns. And indeed they are. Review of financial reports shows that as of 2010, Areva's net debt was €3.7 billion, while EDF's had reached a staggering €42.5 billion. "State within a state" though the French nuclear industry is, the French government must be finding it very hard to cover debts like that.

### Trick 3: Ignore the Quote

Costs increase if design changes are necessary—for example, if the safety regulator requires changes or if the design is not fully worked out before construction starts. That is why plant constructors now aim to get full regulatory approval *before* construction starts, as with the proposed US combined construction and operation licenses, and why they require designs to be as fully worked out as reasonably as possible before construction starts. The risk of design change cannot be entirely removed, especially with new designs where the construction process faces unanticipated problems. Lessons learned while operating reactors might also require changes in the design of others after construction has begun. And, of course, major nuclear accidents (such as Fukushima) necessarily lead to a worldwide review of all plants under construction.

Another fascinating feature of nuclear energy is that huge reactor projects may be launched, often with great fanfare and billions of dollars of subsidies, continued for many years, and then quietly abandoned. The long list of reactors under construction, along with their completion dates, reveals a picture as illusory and miasmic as anything else in the secretive world nuclear business.

The Atucha II site in Argentina, for example, was begun in 1981 and projected to produce a modest 700 MW of electricity. It was supposedly 80 percent complete when the plug was finally pulled on it in 1994, the year when, not coincidentally, Argentina's long-standing nuclear industry was privatized.

## ACTUAL PRICES MAY VARY FROM THOSE ADVERTISED

Again, figures quoted by those with a vested interest in the technology but no actual responsibility for prices—including industry bodies such as the World Nuclear Association and equivalent national bodies—must be viewed with a great deal of skepticism. Prices quoted by international agencies, such as the NEA, are scarcely better, particularly when they are

## RISING CONSTRUCTION COSTS OF NUCLEAR PLANTS

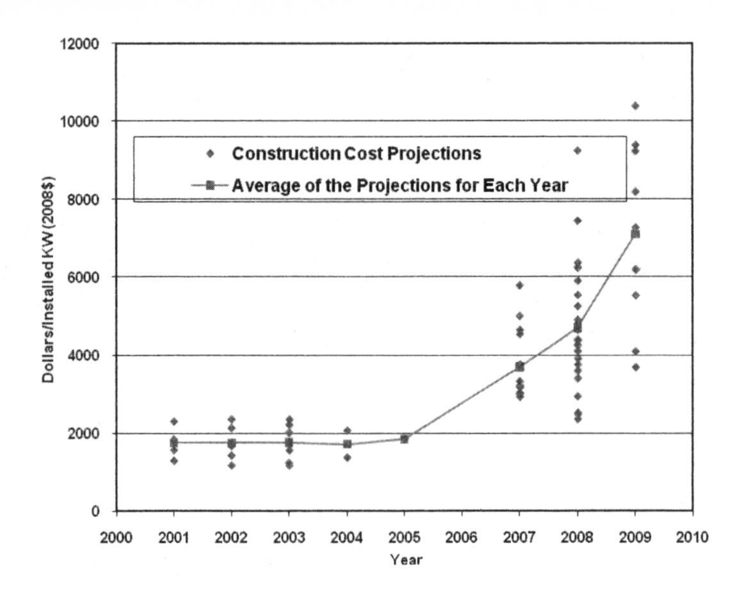


This chart, which reflects some of the most recent nuclear cost projections, shows that estimated construction costs (exclusive of financing) per 1 installed kW production capacity have jumped from a little over $1,000 dollars in 2002 to well over $7,000 in 2009.

*Source:* Based on figure titled "Overnight Capital Costs Projections for New Power Reactors," in Henry Sokoleski, ed., *Global Expansion: Weighing Its Costs and Risks* (Strategic Studies Institute, December 2010); based on chart in Mark Cooper, "The Economics of Nuclear Reactors: Renaissance or Relapse?" Vermont University, Institute for Energy and the Environment, June 2009; www.vermontlaw.edu.

based on indicative, rather than real, costs. Generally, these "costs" are not figures based on actual experience but are estimates provided by national governments, with their own political imperatives and goals.

Apart from political fudges, there are good, practical reasons why costs quoted are not to be taken too literally. As John Fuller, head of General Electric-Hitachi (GEH), remarked to the industry magazine *Nucleonics Week,* site-specific factors can make a significant difference to costs; consider the method of cooling. "GEH had seen plant costs change by $1 billion depending on whether the plant is cooled by saltwater or freshwater."

According to US Energy Information Agency data published in 2011, the most recent construction costs for US nuclear power plants appear to be at least five times the $1,000-per-kW figure that the industry claimed in the late 1990s. There are many reasons for this relentless increase, which is seen in most other large-scale public infrastructure works as well. For example, costs may vary because of a lack of production facilities, which means that utilities hoping to build nuclear plants are taking options on components such as pressure vessels; shortages of the necessary skills as the nuclear workforce ages and is not replaced by younger specialists also cause cost variations.

Certainly, nuclear plants are huge construction projects, on the scale of the Channel Tunnel between France and England or building a highway across the Amazon rain forest. As noted already, such construction projects almost invariably cost far more than predicted. (In China and other developing countries, the costs may be fixed, but the standards slip . . .)

Yet people who sell the projects invariably present the costs as very much fixed and under control. Vendors naturally like to claim that their designs have foreseen all the possible eventualities, since no one wants to buy a plant with unknown complications lurking around the corner.

But nuclear reactors often run into unpredictable factors that necessitate changes to the plan. As discussed, the ambitious new-generation Olkiluoto EPR plant sold by France to Finland is a case in point. After four years of construction, in 2009, the year it was supposed to have gone online, it became clear the design itself was far from complete. In this case, the regulator had particular and serious concerns about the control and instrumentation systems as well as the standard of work, notably concerning welding. The €3 billion reactor, which had been given a new start date of 2011, again had to have the start date pushed back, while the costs continued to rise to at least €5.3 billion.

Overruns can be particularly expensive for nuclear plants. Since Olkiluoto was supposed to be up and running in 2009, its electrical output had already been sold to Finland's industry. Plant operators were thus obliged to buy "replacement power" to supply customers at the prevailing wholesale cost for electricity.

And although new-generation designs sound very promising to political decision makers, from economists' point of view, they increase the risk of cost overruns. Teething problems in the first years of operation will have much more serious implications for a project's long-term viability than maintenance problems in the later years, owing to the logic of costs discounting.

**AMAZING FACT**

It was a sevenfold increase in reactor costs estimates that doomed the so-called Great Bandwagon Market of the 1960s and 1970s, a time when half of planned nuclear reactors had to be abandoned or canceled due to massive cost overruns.

## Trick 4: Special Finance Conditions Available

The largest cost in power from a nuclear power plant is the *cost of capital*, as determined by interest rates and capital charges.

The real (meaning after taking inflation into account) cost of capital varies from country to country and from utility to utility, according to the country risk and the credit rating of the company. (Indeed, in several debt-ridden countries, such as Spain in 2011, nuclear power costs were part of the toxic mix that brought the governments low.) The way the electricity sector is organized is also important. If the sector is a regulated monopoly, the real cost of capital could be as low as 5 to 8 percent, but in a competitive electricity market, it is likely to be at least 15 percent.

When the electricity industry was a monopoly, utilities were guaranteed full *cost recovery*—in other words, they could recover from consumers whatever money they spent. Any investments were very low risk to those providing the capital because consumers were bearing all the risk.

*Does that remind you of anything?* Yes, it is the financial crisis all over again.

Take, for example, the situation in the United Kingdom at the start of the millennium. In 2002 in Britain, about 40 percent of the energy-generating capacity was owned by financially distressed companies (about half of this was the nuclear capacity), and several companies and banks lost billions of pounds on power stations that they had financed.

If there were government guarantees on the market price for power, the cost of capital would be lower, but the guarantees would represent a government subsidy (state aid), and it is not clear they would be acceptable under European Union law. This is where the "carbon price" trick comes in. The chief executive of one company considering building new nuclear power stations, E.ON (UK), told a British newspaper that before it would consider investing in nuclear power, there would have to be some government support to tip the market in its favor—not merely a general tax on its competitors, but a high and guaranteed price for its electricity. Spanish solar power utilities, he suggested, had flourished with prices *four or five times* the going rate; why not the same for nuclear? Dr. Paul Golby said: "We absolutely have to have this change because [otherwise] the current market won't bring forward low-carbon investment."

## Trick 5: Exaggerate Reliability

A good measure of a plant's reliability and effectiveness at producing salable output is its *capacity factor* (also called the *load factor*). Generally, capacity factors are calculated on an annual or a lifetime basis. There can be dispute about the causes of shutdowns or reduced output levels, although, from an economic point of view, this attribution of blame is often of limited relevance.

Amazingly, of all the 132 US nuclear plants built (a figure that is barely half of the 253 originally ordered), 21 percent were permanently and prematurely closed due to reliability or cost problems, while another 27 percent have completely failed for a year or more, at least once.

Unlike construction cost, the capacity factor can be measured precisely and unequivocally. The trade press, such as *Nucleonics Week* and *Nuclear Engineering International,* regularly publishes capacity factor tables. (The capacity factor is calculated as the output in a given period of time expressed as a percentage of the output that would have been produced if the unit had operated uninterrupted at its full-design output level throughout the period concerned.)

Nuclear power plants generally are operated on "base-load" (that is, at a constant and maximally efficient rate), except in the very few countries (e.g., France) where nuclear capacity represents such a high proportion of overall generating capacity that this is not possible.

### OPERATING PERFORMANCE

Vendors and others promoting nuclear technology have assumed that nuclear plants are extremely reliable, with service interruptions due only for maintenance and refueling. Indeed, some reactors are even designed to be refueled continuously and are shut down only for major maintenance, potentially giving capacity factors of 85 to 95 percent. However, as with construction costs, optimistic claims for the reliability of plants have proven hopelessly inaccurate. Historically, performance has been poor. By around 1980, the average capacity factor for all plants worldwide was a lukewarm 60 percent, and the figures had not improved much 25 years later in 2005.*

This reality has a major impact on the economics of nuclear power, an industry where, remember, fixed costs represent two-thirds of the overall cost of power. To the extent that equipment failures cause poor capacity factors, the additional cost of resulting maintenance and repair would further increase the unit cost of power. In addition, a nuclear generator usually has to buy "replacement" power for its customer, often at punitively high prices. As to new reactor designs, they, like new cars, may offer improved reliability in the long term, but also include a greater likelihood of problems in the early years—the period when costs are most difficult for utilities to bear.

* Coal-fired power stations, by comparison, will easily reach 80 percent of their nominal capacity.

## Trick 6: The Low-Running-Costs Trick

Again, given the massive costs of building nuclear reactors, costs to run them often get overlooked in examinations of nuclear economics. Many times it is assumed that if a utility ends up with its loans paid off and a working nuclear reactor, at least, then it can begin to sell electricity at a profit. After all, in these circumstances, the only costs are nonfuel operations and maintenance (or O&M in industry parlance). What is more, the cost of employing staff and maintaining the plant vary little according to the plant's level of output, so the more power that is produced, the lower the O&M cost per MW-hour.

Indeed, the assumption of low running costs has been shown to be wrong several times in the past. In the late 1980s and early 1990s, a small number of US nuclear power plants were shut down because the cost of operating them (and remember, this excludes repaying the fixed costs) was greater than the cost of building and operating a replacement gas-fired plant. Similarly, British Energy, which was essentially *given* its eight nuclear power plants when it was created in 1996, collapsed financially in 2002 because income from plant operation barely covered operating costs.

## Trick 7: Offer to Reuse Old Fuel

Although uranium prices have increased substantially in recent years, the cost of fuel for a nuclear plant remains relatively low as a proportion of all the other costs. In the United States, fuel costs are especially low, arguably because the US government assumes responsibility for disposal of spent fuel in return for a flat fee. Since this fee is based on an arbitrary price set more than two decades ago and not on actual experience—no fuel-disposal facilities exist in the United States or anywhere else—and all the US spent fuel remains in temporary storage pending the construction of a spent-fuel repository, real disposal costs are going to be much higher than projected. (For many years, Yucca Mountain was expected to be the site of the depository, but it will not be, as explained in Myth 6.)

Recently nuclear salespeople have emphasized the potential of plants to run either on waste fuel or on recycled bombs. However, plants that do so lose far more than they might save on fuel and face additional risks and complexities of handling the fuel.

*Reprocessing*—the chemical separation of fissionable plutonium and nuclear fuel—sounds very useful—green, even, a kind of recycling—but

**GREEN NUCLEAR FUEL**

Once a uranium fuel rod has been used (it is then called spent fuel), the uranium will have been contaminated with all kinds of transuranic elements, with exotic names—many of them insanely dangerous, such as plutonium. Of course, some people seek out insanely dangerous things, and plutonium is much in demand by the military for bombs (along with highly enriched uranium).

However, when the by-products are not being used to kill people, they have to be dissipated. When this is done with water, the process produces vast quantities of contaminated water (of the kind that the Japanese have been releasing into the Pacific following Fukushima). Alternatively, higher-level waste can be mixed with lower-level waste, or neutron-absorbing elements, such as boron, or with cheap materials, such as sand. But whatever method is used, the end result is always a greater volume of waste. It is veritably the Sorcerer's Apprentice syndrome!*

* "The Sorcerer's Apprentice" is a poem by Goethe in which an old sorcerer departs his workshop, leaving his apprentice with chores to perform. Tired of fetching water by pail, the apprentice enchants a broom to do the work for him—using magic in which he is not yet fully trained. The floor is soon awash with water, and the apprentice realizes that he cannot stop the broom because he does not know how.

it is expensive and, unless the plutonium produced can be used profitably, it does nothing to solve the problem of total waste disposal. In fact, it creates additional waste.

There are only two commercial reprocessing plants in the world—Sellafield in the United Kingdom and Cogema in France. These plants take in used fuel rods from plants around the world and dissolve them in hot nitric acid in order to recover usable uranium. Eventually, the plants produce small amounts of plutonium and some more highly radioactive waste. For commercial nuclear energy companies, the only practical use for plutonium is to mix it with uranium to produce mixed-oxide fuel: MOX. But even by the standards of nuclear power, MOX is a dirty and dangerous fuel.

Despite its reassuring name, reprocessing merely splits the spent fuel into different parts and does not reduce the amount of radioactivity that must be dealt with. In the process it creates a large amount of low- and intermediate-level waste, because all the equipment and material used in reprocessing becomes radioactive waste. The small amount of highly radioactive waste it produces is turned into a powder and mixed with glass to produce pellets. Recall that by international agreement, all countries are responsible for their own nuclear waste. Indeed, the French and British reprocessing centers include in their contracts for international customers the obligation to accept back their own waste. But—and here is another nice nuclear trick—the return date is never specified.

## Trick 8: The Bonus Lifetime Extension

There is a movement afoot to extend the life of existing plants, and pressurized water reactors are now often expected to be run for more than 40 years, compared to their design life of, say, 30 years. Since all the capital costs have long been paid off, it might be expected that at least in their twilight years, nuclear plants might begin to make money. Alas, no. Life extensions require significant new expenditures to replace worn-out equipment and to bring plants up to current safety standards.

And although these new-generation plants promise to run for longer—60 years instead of 40—little money is saved, since commercial loans typically have to be repaid in the first 15 to 20 years. So there is not even the potential for cheap electricity once capital costs have been repaid.

## Trick 9: The Discounted Costs Trick

Discounting costs is a piece of financial engineering every bit as audacious as splitting the atom. Essentially, a small sum of money is put aside to cover future costs. The claim is that, due to the magic of compound interest, this small amount will grow and become sufficient to cover even enormous liabilities. At a discount rate of 15 percent, costs or benefits more than 15 years forward have a negligible value in conventional economic analysis. This is why it is possible for bodies like the World Nuclear Agency to declare on its website and in its pamphlets that "nuclear energy fully accounts for its waste disposal and decommissioning costs in financial evaluations." (The weasel word here is "accounts.")

Steve Thomas again:

> [T]here is a significant mismatch between the interests of commercial concerns and society in general. Huge costs that will only be incurred far in the future have little weight in commercial decisions because such costs are "discounted." This means that waste disposal costs and decommissioning costs, which are at present no more than ill-supported guesses, are of little interest to commercial companies.

Costs arising more than, say, ten years in the future have little weight in an evaluation of the economics of a nuclear power plant. Very conveniently, the time from placing a reactor order to the completion of decommissioning could span more than 200 years. By the magic of the financial strategy, very large decommissioning costs have little impact on balance sheets, even with a very low discount rate that is consistent with investing funds in a very secure place. If, for example, a British magnox plant* will cost about $1.8 billion to decommission and the final stage accounts for 65 percent of the total (undiscounted) cost ($1.2 billion), a paltry sum of only $28 million invested when the plant is closed will have grown sufficiently to pay for the final stage of decommissioning.

Problems arise if the cost has been initially underestimated, the funds are lost, or the company collapses before the plant completes its expected

* Magnox nuclear power reactors were originally designed in the United Kingdom where they are still in use, and then exported to other countries, both as a power plant, and, when operated accordingly, as a producer of plutonium for nuclear weapons. The name magnox comes from the alloy used to clad the fuel rods inside the reactor. Magnox reactors are pressurized, carbon dioxide–cooled, graphite-moderated reactors using natural uranium (i.e., unenriched) as fuel and magnox alloy as fuel cladding.

lifetime. Britain has suffered all of these problems. The expected decommissioning cost has gone up severalfold in real terms over the past couple of decades. Not bad enough? In 1990, when the Central Electricity Generating Board (CEGB), which owned power stations in England and Wales, was privatized, the accounting provisions extracted from consumers over the years in their electricity bills were not passed on to the successor company, Nuclear Electric. As Steve Thomas says, the subsidy that applied from 1990 to 1996, described by Michael Heseltine, sometime UK energy minister, as being to "decommission old, unsafe nuclear plants," was in fact spent as cash flow by the company owning the plant, and even the unspent portion has now been absorbed by the Treasury.

## Trick 10: Abolish Liability

As of September 2001, liability limits for the countries that are members of the Organization for Economic Cooperation and Development (OECD) showed that there were remarkable variations in opinion worldwide about both the risks and the consequences of nuclear accidents.

How much insurance cover does a nuclear plant actually need? No one seems to know. As a glance at the table shows, the extraordinary range is from a rather trivial €12 million in Mexico (of course any accident there would not affect, say, the United States!) to unlimited liability in Germany along with a fairly serious-looking €2.5 billion liability fund. Could it be that German technology is 200 times inferior to Mexican? Or could it be that there is a lack of an international regulation of the nuclear industry? After the Chernobyl accident, when the US General Accounting Office conducted an analysis of the off-site financial consequences of a major nuclear accident for all the nuclear power plants then operating in the United States, it also found the estimates ranged widely—from a paltry $67 million to a bankruptingly high $16 billion.

In hindsight, it is curious that Japanese nuclear companies also have unlimited liability. Since its flagship nuclear company, TEPCO, now has a major disaster to pay for, one might expect its share price to drop and for

## LIABILITY LIMITS FOR THE OECD COUNTRIES AS OF SEPTEMBER 2001

| Country | Liability limits under national legislation[a] | Financial security requirements[a, b] |
|---|---|---|
| Belgium | 298 mln € | |
| Finland | 250 mln € | |
| France | 92 mln € | |
| Germany | unlimited | 2,500 mln €[c] |
| Great Britain | 227 mln € | |
| Netherlands | 340 mln€ | |
| Spain | 150 mln€ | |
| Switzerland | unlimited | 674 mln € |
| Slovakia | 47 mln € | |
| Czech Republic | 177mln € | |
| Hungary | 143mln € | |
| Canada | 54 mln € | |
| United States | 10,937 mln € | 226 mln € |
| Mexico | 12 mln € | |
| Japan | unlimited | 538 mln € |
| Korea | 4,293 mln € | |

*Source:* Unofficial Statistics—OECD/NEA, Legal Affairs

*Notes:* [a] using official exchange rates from 06/2001 to 06/2002, [b] if different than the liability limit, [c] 256 mln € insurance, 2.5 bln € operator's pool, 179 mln € from Brussels amendment to Paris Convention.

the company to disappear. Indeed, after the accident, its share prices did start to plummet, but then a mystery buyer appeared on the exchanges—evidently a believer in the ability of the electrical utility to find several billions of dollars of extra income—purchasing shares voraciously and thus putting a floor under the company's share price. At the time of writing, no one knows who that buyer might have been. But could it have been state actors, perhaps even the Japanese government itself, rather than private investors? Of course such a thing would be disgraceful, as it might encourage nuclear companies to imagine that they can enjoy the profits and offload the problems and crises onto taxpayers. However, as long as governments step in to underwrite the cleanup costs after an accident, the nuclear industries themselves have a very rosy future: They can apply their expertise to the potentially much more lucrative business of entombing their reactors and decontaminating the countryside.

Anyway, at present, the liability of plant owners is limited by international treaty to only a small fraction of the likely costs of a major nuclear accident. The Vienna Treaty, passed in 1963 and amended in 1997, limits a nuclear operator's liability to 300 million Special Drawing Rights. The British government underwrites residual risk beyond £140 million, though the limit is expected to rise under the Paris and Brussels Conventions to €700 million (£600 million, or getting on for a good $1 billion). Governments agreed, at the time the treaty was passed originally, that the limit on liability was essential to allow the development of nuclear power, but this cheap insurance serves, in practice, as a large subsidy.

The sheer scale of the costs caused by Chernobyl and Fukushima, which are acknowledged to run to hundreds of billions of euros, means that a major accident would bankrupt insurance companies. It is more than fortunate, then, that the companies are absolved of such risks by international treaties that shift the risks to taxpayers. But this subsidy may be less significant than it seems, because, amazingly, *less than half the world's nuclear reactors are covered by any of the existing international agreements.*

## Trick 11: The Decommissioning Trick

The decommissioning trick is very simple: Do not do it. This can be stated quite publicly—for example, under UK plans, plant owners need not begin the key—and most expensive—stage of decommissioning until 135 years after plant closure. If the public was irritated to see irresponsible bankers and speculators get away scot-free with their risky business practices during the banking crash, it remains remarkably sanguine about the prospect of nuclear financiers evading the consequences of their activities. The delay in decommissioning is, however, vital to the viability of nuclear finance. By putting costs off for a century or so, even very large decommissioning costs impact the balance sheet only slightly, and even then with a *very* modest assumption about the interest paid on money put aside for the eventual decommissioning, such as 3 percent. Using this kind of accounting convention, it can be calculated that for a typical British magnox plant, of which the main stage of decommissioning is conservatively expected to cost about $1.2 billion (were it undertaken today), the paltry sum of $28 million will suffice to cover the future liabilities, as long as it is invested when the plant is closed. The idea is that the nest egg will grow over the intervening century or so sufficiently to bridge the difference.

Now that is magical thinking. Mind you, in the best spirit of conjuring tricks, even such paltry sums put aside by nuclear companies to cover future liabilities have a habit of disappearing. When the CEGB was privatized, about £1.7 billion (perhaps $2.5 billion at the time) carefully set aside for the future was included in the offer. But when the company was sold for a song and dance, reflecting investors' dismal assessment of the costs of nuclear electricity, effectively two-thirds of the set-aside funds also disappeared. (Just to complete the trick, the British government did not pass on *any* of the sale proceeds to the company that bought the nuclear power plants, thereby magicking away the remainder of the funds.)

## Trick 12: Transfer the Costs—in Time and in Space

Another great source of confusion in the economics of nuclear power is the ubiquitous use of transfer pricing. This is the technical term for fudging and shuffling costs across time and over national frontiers (or within a country), and it is a stock-in-trade not only for the nuclear industry but for other not-so-clean energy industries, notably coal mining and oil refining.

What these industries do is transfer profits and costs from one physical locality or operating time period to some other, preferably distant, place or time—usually as far in the future as possible. The result is pricing down the costs and pricing up the profits—or, in technical finance terms, "diluting costs and valorizing profits."

All this is, of course, to the delight of experts who are given the opportunity to juggle so many variables and play with so many different criteria. Naturally, they always manage to deliver the answer that their money men want to hear: Nuclear power is the cheapest and most efficient form of energy available.

Nuclear pricing transfers operate across geographical space in the case of major uranium-exporting countries such as Niger. In the process of extracting and processing uranium, the French-owned Areva conglomerate mines consume almost one-third of that entire country's (admittedly modest) total national electricity production. This electricity, climate change campaigners for nuclear ought to note, is mostly produced from oil, a fact that immediately dents nuclear power's carbon-free image.

Niger then exports this uranium, by oil-fueled trucks and oil-fueled bulk cargo ships, to run the climate-friendly nuclear power stations of France, Japan, and other countries. As a host of reports by environment and human health–oriented nongovernmental organizations show, Niger's Faustian pact exchanges short-term revenue benefits for its ruling elite and their immediate successors with heavy costs for the natural environment and local communities—costs that will be paid for many decades to come.

## THE BOTTOM LINE

Behind all the financial camouflage, what is the answer? How much does nuclear electricity *really* cost?

On one hand, the methods used for calculating the costs of producing electricity, and therefore the minimum price at which it can be supplied, should follow the rules of classical economics—taking account of capital, fuel, labor, and operating costs—and should therefore be quite transparent. On the other hand, many of the factors determining the price consumers actually pay are highly political. In the past, governments often sought to lower electricity costs artificially, but these days consumers may be charged extra to support "green energy" or the cleanup costs of past nuclear ventures (as British consumers do). For example, the typical bulk electricity prices in the most developed electricity trading markets of the United States, and the even bigger markets of western and eastern Europe, can be as low as 3.5 or 4.5 US cents per kw/h, yet the final consumer price varies hugely, for a huge variety of reasons and factors.

In fact, the question, which looks simple, conceals many assumptions. The first is the question of which kinds of energy count.

As discussed in Myth 1, noncommercial energy sources, led by wood and dung, provide almost as much energy to the world as commercial nuclear power—that is, about 625 million tons oil equivalent in 2009. Political elites do not give world consumption of fuelwood and dung much attention, not merely because of their unfashionable connotations and image but also because a huge proportion—likely more than 90 percent—of such energy has no cash value at all.

The kind of energy that comes into people's houses via wires is very easy to both measure and price. Add to which, when you plug in an electrical appliance, you are really not concerned if the power came from coal, natural gas, landfill methane, solar cells, wind farms, hydropower—or nuclear power. Often the debate over the role of nuclear in a country's energy portfolio is artificially circumscribed to suppose that this tidy kind of electrical energy is all that we need think about.

As the tables in Myth 1 show, nuclear energy supplied only about 6 percent of the world's secondary energy in 2009. The same year, it also supplied about 15 percent of the world's *electricity.* But if we focused only on pure primary energy, we would find that nuclear electricity was equal to no more than around 2.5 percent of the world's total.

**AMAZING FACT: THERMAL OUTPUT AND COST**

At least no mystery complicates our access to engineering-type industry and international comparative data on thermal electric power plant costs and economic performance. The US Energy Information Agency, for example, produces a yearly report on electric utility capital and operating costs in its Annual Energy Outlook series. In its 189-page report, issued in November 2010, the capital costs for nuclear plants located in the United States were estimated as rising by 37 percent for a standard US pressurized water reactor. In simple terms, the cost of building nuclear power plants is doubling every two years.

Put another way, we can answer how much nuclear power really costs by looking at the ratio between total input costs and the total value of outputs during a plant's lifetime. The cost side should include all the external costs, such as human health and environmental costs, the costs of dismantling the plant, and all restoration and reclamation costs for the power plant site. It should include the cost of cleaning up mining waste or spill zones—zones that can cover whole regions and linger on long after the end of activities. For every pound of uranium ore, many tons of rock must be crushed, for example, and the uranium extracted using acid solution. (Since even the richer ore bodies worked in most uranium mines contain less than 1,000 parts uranium per million, it follows that each ton of uranium extracted leaves 1,000 tons of mine waste behind.)

This principle applies with full force for coal and lignite power plants, but such postmortem costs are brazenly pushed aside for nuclear plants.

The nuclear power plant's cash inputs include building the plant and dismantling it when it is taken out of service; supplying its fuel, labor, building and other materials; and its maintenance and services. The plant's output benefits are mainly, often only, the value of electricity, although in a few cases they may also include heat energy output from the plant, perhaps for desalination. Both forms of energy also produce tax revenues for local and central governments and profits for the operating company and its shareholders.

Actually, one economic argument *for* nuclear power is that, in addition to jobs created by the power plant in its operating lifetime, considerable numbers of jobs also can be generated by restoration and reclamation of the site, environmental protection, waste management, and other economic activity after the end of its life. For nuclear power, this job creation potential is very large, but, unsurprisingly, the nuclear industry and its apologists are not eager to draw attention to the necessity of such tasks.

A related Faustian bargain applies to many local governments in the developed world. Cooperative communities that accept nuclear reactors (or waste repositories) benefit from wonderful streams of nuclear money. Schools, community centers, sports facilities sprout up like mushrooms around nuclear plants. (The same effect explains why, for instance, the tiny and isolated villages around the Dounreay plant and waste repository in Scotland have new pews in their village churches.) The balance sheets of local counties are transformed. But soon enough, the money dries up, and the communities are left with depleted employment and environmental resources and a blighted long-term future.

Uranium mining in Niger is just one example of the way in which western transnational businesses can transfer the benefits away from the geographical source while leaving the dirt behind. However, the most dramatic form of nuclear transfer pricing operates in the time dimension. It includes previous hard-cash subsidies to the nuclear industry upstream and reactor dismantling and decommissioning downstream.

No one can give exact figures for how much money the civil-nuclear system has received, often starting five decades ago, and goes on receiving today from the military-nuclear complex. These subsidies are so vast that any figures offered are always controversial. They are nearly impossible to get because anything military, especially nuclear military, is classified, but estimates for these subsidies from the atom bomb era in France, compiled by *Reseau Sortir du Nucleaire* (Nuclear Phase-out), go as high as $250 billion in today's money over the period starting about 1950 and ending around 1968. France, in this sense, is typical of the nuclear powers. The subsidies came in the shape of complete nuclear infrastructures, transferred from the military side to civil reactor operation, but they were never directly *charged to* the civil nuclear power industry.

The physical scope of these military-cum-civil nuclear infrastructures, in many cases more than half a century old, is immense. It spans the entire nuclear system, from fuel production and reprocessing to waste disposal and reactor decommissioning. Here we find another key example of the strange timelessness of nuclear costs that are never borne by the industry. Regarding early reactors, whose status was never firmly defined as military, experimental, or civil, there are multiple examples in all the old-nuclear countries of dismantling and site remediation works that have already taken 25 years and can easily continue another 15 to 25 years.

These seemingly endless works have cost governments and taxpayers—but never the power plant operators—hundreds of millions, and sometimes billions, of dollars to date.

For nuclear waste, which requires storage for at least 10,000 years, the costs are extremely difficult to estimate. Because they are so enormous, they are quickly given future value write-downs that look good on paper. What initially costs about $100 million a year—for example, the first ten years of operations in the now-abandoned US Yucca Mountain project or the proposed but "not yet" built Bure project in France—will magically slim down to a handful of small change.

That is, of course, on paper—even if the amount of waste being stored has massively grown in that period, along with the real costs. Such is the nature of creative accounting. The same trick, in the opposite direction, is used frequently by the nuclear lobby and its climate change allies to inflate the future costs of not mitigating what they claim to be the possible economic damage from global warming. For example, a report by Lord Stern for the British government exaggerated the future costs of global warming in some scenarios by using high annual interest charges, which purported to show the costs of not tackling global warming rising to $68 trillion a year by the middle of this century.

There is no real difference between nuclear-civil and nuclear-military activities, but since the second concerns national security, its costs can be kept secret, which is the next best thing to not having any.

Today, many countries are stuck with the massive, and massively costly, legacy of the attempt to make nuclear energy commercial by throwing public money at it. As one pressure group, the Union of Concerned Scientists, put it in a 2009 report on the history of nuclear power in the United States: "More than 30 subsidies have supported every stage of the nuclear fuel cycle from uranium mining to long-term waste storage." The group concluded that from start to finish, right through the value chain, nuclear power needed—and still needs—subsidies.

"Added together, these subsidies have often exceeded the average market price for the [electric] power produced . . . in some cases it would have cost taxpayers less to simply buy the kilowatts on the open market and give them away."

Even that stalwart supporter of the industry, *The Economist,* observed back in 2001 that "[n]uclear power, once claimed to be too cheap to meter, is now too costly to matter—cheap to run but very expensive to build."

# MYTH 5

# NUCLEAR POWER TRUMPS GEOPOLITICS

*For many countries, nuclear science is the ultimate trump card. In this archive picture, the impressive "Jumbo" atomic device is being positioned at the Trinity test site at Alamogordo, New Mexico, in 1945, where the first-ever atomic bomb was exploded. Actually, although Jumbo looks like a fearsome weapon, it was in reality a hollow sham—a mere steel container constructed to help conserve the precious plutonium in the eventuality that the actual nuclear device placed inside it failed to go off.*

*"Jumbo," yet another example of how appearances in the nuclear world can be deceptive. (US National Archives)*

**ASK SOMEONE TO LIST RECENT NUCLEAR CRISES AND** they probably won't mention Westinghouse, in 1978. But of them all—Windscale, Chernobyl, Three Mile Island, Fukushima—it is in many ways the most revealing.

In 1978, the US corporation Westinghouse was the undisputed number one in world civil nuclear energy. But even in its heyday, its finances were crumbling behind the facade. The cause of the company's woes was uranium shortage or high prices for uranium—which is not a semantic difference. If anyone still believes there is no shortage of uranium fuel for the world's reactors, then after Westinghouse they have to accept that there is a real problem about how much it costs.

What happened to Westinghouse in the three years preceding 1978 is simple to explain. Uranium prices increased very fast, and Westinghouse was unable to get the element at the prices it had forecast and written into its supply contracts, which stipulated an unrealistically low price for the fuel. And although Westinghouse weathered the storm, it emerged with battered credibility into a market for reactors that was contracting fast. Ironically, if logically in market terms, the acute shortage of uranium in the period from 1974 to 1978 then turned into a glut, with supplies exceeding demand until at least 1985. Prices, of course, crashed.

In fact, uranium prices have always been volatile. During the race for the bomb by the five permanent member nations of the United Nations Security Council during the 1950s and 1960s, uranium prices rose to well above their recent peak, in 2007. In today's dollars, prices often hit $350 per kilogram, but the amounts needed then were tiny compared with the needs of today's 440-strong world civil reactor fleet.

In 2011 dollars, the 1970s crisis was mapped by per-kilogram prices rising from around $80 in 1974 to more than $250 in 1977, before crashing. As late as the year 2000, uranium prices were less than $18 per kilogram, but they reached nearly $300 a kilo in 2007, before they crashed again in 2008–2009. Prices then rose to about $140 per kilogram in early March 2011. Then the Japanese crisis hit, sending prices back down again.

Uranium mining, supply, and pricing—through a totally closed company-to-company supply system that cannot be called a true "market"—is extremely cyclical. There are guaranteed bursts of high prices and guaranteed periods of low prices.

Of course, it is true that the major expense of running a nuclear reactor, as explained in Myth 4, is paying the bills for building the reactor. Nevertheless, fueling a newly completed industry-standard 900-megawatt (MW) reactor that needs about 200 to 250 tons of uranium for its initial core loading costs an amount, which if somewhat unpredictable, is always large.

Supporters of nuclear power always wave aside variations in the price of uranium, because even if uranium is not exactly an abundant mineral (its crustal abundance of 4 parts per million ranks it low among world minerals), world reactor fuel needs, even in 2011, are only about 68,000 tons. That is a lot compared to gold production, anticipated in 2011 to be about 2,500 tons, but compared to coal, whose power stations will probably want feeding with some 7 *billion* tons the same year, it is a speck of sand. The nuclear industry thinks that as long as uranium needs stay somewhere down near the gold end of the spectrum, particularly when compared to the extreme high of coal, all will be well. Of course, it is also physically practical for countries to stockpile several years' supply of uranium to feed their reactors, if need be.

This cozy argument, opposing what is sometimes called the threat of peak uranium, contends that uranium resource depletion or higher uranium prices, and growing numbers of new reactors pushing demand ever higher, will not generate an endless series of financial crises for nuclear power. It is argued that the uranium supply shock of the 1970s that tipped plant builders, such as Westinghouse, into a frenetic bidding war with the so-called financial community that the plant builders lost was simply a one-time thing. The industry insists that it is much better equipped to manage temporary shortfalls today.

But is it? The negative proof starts with the persistently high uranium spot prices, even post-Fukushima, that have made uranium mining equities, venture capital, and merger or buy-out operations strong assets for the hedge fund community. These investors have done the math and discovered that uranium supply is at least as sensitive to supply and demand constraints as the world oil market. The industry's counterarguments are more of the same: Only small tonnages of uranium are needed, uranium supply cutoffs are unlikely, and any shortfall will only be temporary, perhaps due to uranium mine expansion not going ahead as fast as is hoped or planned.

What is more, the industry argues that new technology could cut uranium needs or might totally replace uranium as the main fuel needed for the expanding world reactor fleet.

But the truth is that mine supply in the year 2010 was less than 55,000 tons. A critical and fragile list of stopgaps made up the 20 percent, or 13,000-ton, shortfall. These stopgaps range from stocks held by uranium mining firms, reactor builders, and fuel fabrication companies; mixed oxide (MOX) recycled fuel; and the US-Russian Megatons to Megawatts program of dismantling and recycling warheads materials from atomic weapons. All of these sources are shrouded by commercial or state secrecy, but as any uranium-savvy hedge fund analyst knows, the nuclear industry faces almost permanent shortage. For years into the future, the uranium market will always be tight. The secure, reliable uranium supplies of nuclear myth making and the 50-fold increase in nuclear generation dreamed of by the climate change campaigners simply do not exist.

To make matters worse, in terms of geopolitics, today's world uranium supply increasingly comes from outside the Organization for Economic Cooperation and Development (OECD) group of countries and its old-nuclear nations, making nuclear power's vaunted energy security claim just one more myth. Africa and Central Asia are becoming the key supplier regions, supplanting even the substantial production activity in

### FROM MEGATONS TO MEGAWATTS

What better way to try to show how military uses of nuclear materials can and are being transmuted into peaceful ones than the story regularly replayed in the media of the Megatons to Megawatts program in the United States? Recycling uranium, not so much turning swords into plowshares but bombs into electrical lighting, is the basis of the program. In recent years, the major authorized company operating this market, Cameco, which is the world's largest uranium mining and fuel supply company, was said to be buying and reselling around 7 million pounds (over 3,000 tons) of Russian ex-military uranium each year.

For a select group of North American companies, but above all for Cameco itself, for which this supply represents about one-quarter of its total sales of uranium, the program has been very profitable. For the 100 or so operators of civil nuclear reactors in the United States, it met almost half of their annual fuel needs, at an advantageous price.

However, at the time of writing, the Megatons to Megawatts program is scheduled to stop in 2013. Which is where geopolitics reenters the picture. Cameco's agreement with the sole Russian supplier, the state firm Techsnabexport (Tenex), will be renewed only if the US administration and the Russian president—that is, Barack Obama and the Medvedev-Putin duo—agree to another round of US cash for out-of-date Russian warheads. Quite apart from US-Russian relations being regularly subject to periods in the deep freeze, this short-term fix can hardly be repeated indefinitely—presumably the supply of stockpiled warheads is finite—so future supplies from this source are far from certain.

Canada and Australia, bound hand and foot as they are by adherence to environmental standards, health and safety norms, and legal codes.

The industry response to this rising threat of both undersupply and dependence on geopolitically risky sources typically consists of new-tech quick fixes. Two in particular are MOX fuel, which contains high amounts of plutonium and other deadly radionuclides recycled or reprocessed from spent fuel and similar fuel materials from dismantled US and

Soviet warheads. Both these sources come with question marks relating to their future supply: The Fukushima disaster highlighted the dangers of MOX fuel, with at least one of the six stricken reactors containing MOX. This made emergency operations all the more complex and hazardous, and both the United States and Russia are running out of surplus (or past their blow-up-by date) warheads to cannibalize and recycle for reactor fuel materials, with the crunch expected to come about 2015.

Other technical fixes include a stunning array of Rube Goldberg/ Heath Robinson ideas, ranging from the fuel side to the design, form, type, and utilization of new-generation reactors. Among these is the plutonium economy, where fast breeder reactors would be built in massive quantities, with immense quantities of plutonium being "bred" from the spent uranium fuel of current-generation or classic reactors. Another quick fix would be simply to replace all current uranium-dependent reactors with thorium-fueled ones, as thorium is much more abundant in the earth than uranium.

Alas, these remain near-fantasy solutions, not simply because of the fantastic costs but also because of temporal, technological, and industrial constraints. In the real world of today's uranium-fueled reactors and of those optimistically being planned with more parsimonious appetites, nuclear electricity still requires a diet of fresh-mined fuel. This fuel is usually uranium nitrate—the infamous yellowcake parodied in the cult 1980s comic *The Lone Prospector.*

Apart from crunching up mountains there are only a few other sources, which are simple to list. Power companies, reactor manufacturers, national agencies, and the military hold stocks of uranium, and there is the MOX, which is produced by mixing fresh uranium, or recycled uranium from used, or spent, fuel rods, with plutonium and other transuranic elements also separated from spent fuel rods. MOX is a highly dangerous and chemically very toxic fuel, at present commercially produced only in France and the United Kingdom, with customers in Japan.

These very real shortfalls cast doubt on the grand strategies of the nuclear industry to replace fossil fuels as the world's basic energy source.

### MAD MOX

US government websites and the pronuclear World Nuclear Association claim that, in 2009, weapons-grade uranium from Russian warheads converted to reactor-grade fuel met as much as 45 percent of the total reactor fuel needs of the United States, a figure that is approximately equal to one-eighth of world total reactor fuel needs. The US Enrichment Corp., the sole US state-founded company empowered to handle these materials, proudly announced that "the equivalent" of 15,000 atomic warheads had been converted to fuel from the start of the program in 1993 to the end of 2009.

However, amounts and tonnages of the uranium-substitute fuel produced by MOX processing and by the Megatons to Megawatts program are disputed and controversial. Even on the most optimistic assumptions, MOX fuels cover about 2,500 to 4,000 tons per year of uranium needs, and the Megatons to Megawatts program covers about another 8,000 tons per year, all of which still falls far short of the deficit of uranium sourced from mines relative to world reactor needs—about 13,000 to 14,000 tons per year in recent years.

Consider, for example, the ongoing geopolitics of the Mad Quest for Yellowcake.

Gold is the first "yellow metal" that sent prospectors out to the Wild West eagerly searching for nuggets, but now we have yellowcake, the raw form of uranium present in similarly tiny quantities in rocks. Unlike gold, however, there is a little bit of uranium in almost all igneous rocks—indeed, as recently as the 1980s, France extracted as much as 8,000 tons of uranium a year, mainly from pulverizing its Brittany landscape. Since, however, uranium mining produces huge amounts of waste and releases radiation and toxic chemicals into the environment, now there is a dash for Africa and potential new sources in poor countries, such as Afghanistan and Mongolia. In this way, nuclear energy has spawned a new wave of imperialist wars, just as "black gold"—oil—did before it.

Despite this murky underside to uranium supplies, there is a particularly persistent myth that nuclear energy will provide energy security and that, thanks to the atom, countries are shielded from supplier blackmail and high prices. It suggests an alternate universe in which countries without their own fossil fuels can still achieve energy independence, simply by buying shiny new nuclear reactors. In this parallel world, nuclear energy is essentially limitless, based on "wildly abundant" uranium, cheap, clean (low carbon dioxide), high-tech, modern, and has nothing to do with bombs. The dull reality that nuclear electricity everywhere in the world could be replaced, if need be, by either importing electricity from neighboring countries (as of the winter of 2010–11, even France was doing this) or by the unpopular and unprofitable option of using a bit less, is completely lost. Actually, the Japanese did implement the latter very successfully, perhaps even fanatically, following the Fukushima crisis and the forced shutdown of much of its nuclear capacity. This success must be contrasted with what happens when supplies of oil and gas are disrupted, as has happened several winters running in Europe owing to the geopolitics of the Middle East and Russia. In such cases, transport grinds to a halt; hospitals and schools close; and countries immediately feel the cold.

No wonder, then, that from 1995 to 2000, with world reactor orders slowly recovering from the long nuclear winter that started in the early 1980s, the nuclear industry and its supporters in government quickly linked fears of global warming and rising oil prices to intensify public support for nuclear power as the silver-bullet solution to these twin evils.

The former sales pitch worked best, at least for a few years, in the old-nuclear countries of the OECD, especially with politicians and the corporate elite; but today, it is the issue of rising oil prices that has real traction with the leaderships of the emerging and developing countries.

The threat of future oil supply crises is one reason why, with China and India leading the way, as many as 20 developing countries, from Ghana and Sudan to nearly all major Arab states and through Asia, from Mongolia to Bangladesh (like so many naive shoppers seduced by unlimited in-store credit), have stampeded into a nuclear finance trap. The first element in the changing fortunes of the nuclear industry is well known: the earthquake, tsunami, and subsequent meltdown of Japan's nuclear industry. But the other part is more painful still for the developing countries: The Arab Spring and the gathering storm clouds for world monetary and financial stability have pulled the rug out from under developing economies and their debt-laden plans for ever larger, ever grander reactor-building projects. Thus, this is a trap which has already slammed shut on many of their helpless peoples.

As explained in more detail in Myth 8, the massive recentering of the nuclear industry sales effort to the South enabled a potential new and massive asset bubble that was intrinsically unsafe and surely doomed to failure. Surveying the scene, as of 2011, the chances of the new nuclear asset market recovering and rebounding are hard to call, but its beguiling, complex, and fragile financial packages are likely the real reason why the nuclear sales success in the South has already begun to unravel. In many cases, these nuclear plans came with 30- to 50-year credit-based financing packages, which, unless the countries maintain the recent strong growth of their economies and national budget and trade surpluses, will turn into millstones of debt. As to the wealthy nuclear clients, and the key three oil exporter countries of the Arab world—Saudi Arabia, the United Arab Emirates, and Kuwait (all of which have signed contracts for massive reactor projects since 2009, backed by credit-based floating financial packages using tradable instruments)—the claim by their leaderships is that nuclear power will somehow stretch the oil age by reducing domestic demand and enabling these countries to keep exporting oil longer, even as their own oil production diminishes. However, for many observers, this sudden enthusiasm for nuclear power is better explained by fear and loathing of nuclear Iran, just across the straits of the Persian Gulf.

But for most of the new-nuclear countries, many of which suffered heavy economic damage for as long as 15 years after 1985, during the Third World debt crisis, nuclear financial packages are a toxic bomb that indeed trumps geopolitics. For these countries, the long-term sustainability of the financial packages, uncertain performance under stress-test scenarios of falling economic growth, lower export revenues, budget and trade deficits, weakening national currencies, and so on can point to only one destination: mountains of unpayable debt.

We've been there before. In fact, even as long ago as 1975 to 1979, nuclear power financing had already become a classic asset bubble. Nuclear power, for all intents and purposes, lost its credibility at that time—not due to health, safety, or environmental costs and risks but due to fantastic inflation in the nuclear sector. Right through the value chain, starting with a spiral of building and operating costs for plants, nuclear power had become an overheated asset bubble and imploded.

And it was the 1979 Three Mile Island incident—in which, as the nuclear industry likes to say, "no one died"—that burst the bubble. After the infamous American power station malfunctioned, some six years of patient and extremely expensive operations were required before the reactor core could be reached safely and worked on. This revealed to investors the cold economic and financial reality: Nuclear energy is incredibly expensive as well as fantastically dangerous.

In recent years, as part of the nuclear renaissance, the industry has tried to reinvent itself as "energy without frontiers." Nuclear power was promised as a way around the complex web of world energy politics; all any country needed to do was buy a nuclear reactor, fill it up with uranium fuel, and then sit back and laugh at its neighbors struggling with energy price fluctuations and oil shocks. Nuclear energy, in the minds of many governments, was inherently "independent" energy.

On the contrary, the reality is that nuclear energy is one of the most highly politicized and manipulated forms of energy supply in the world. To demonstrate this, it suffices only to take, a brief "Whistle-stop Tour" of the world, stopping at each of the actual or proposed nuclear marketplaces. On the way, we will find always the same handful of transnational corporations, operating within spheres of influence every bit as defined as those of the weapons industries and always seeking to maximize their profits and interests irrespective of long-term consequences, let alone their clients' energy needs. For all that the United Nations Nuclear Suppliers

Group boasts of having as many as 46 member nations, only a handful of companies are actually active in the field: Areva of France, Westinghouse-Toshiba and GE-Hitachi of the United States and Japan, KEPCO of South Korea, and RusAtom of Russia.

## A TOUR OF THE WORLD OF NUCLEAR POWER

### Argentina

Argentina, always a nuclear enthusiast, was one of the first countries to build a nuclear infrastructure in the 1950s benefiting from the émigré expertise of its German experts, some with Nazi links. The government even set up a special nuclear research institute under one such, Ronald Richter, focusing in the first years on research and different nonenergy nuclear applications. However, as far as generating electricity goes, Argentina has not progressed very far. Apart from two reactors producing negligible quantities of very expensive electricity, it has abandoned a third, more serious 750 MW reactor called Atucha II. As mentioned in Myth 4, this plant was begun in 1981, but lack of financing halted production several times until it was abandoned indefinitely in 1994, the same year nuclear energy production at the existing two plants was transferred to Nucleoeléctrica Argentina SA, a limited company. This was part of a government effort to drastically cut public spending as well as to privatize and open up markets for high-tech. However, dreams of selling shares to private investors failed amid serious doubts about profitability. Today the company is still owned directly by the national ministry of energy.

In 2006, under the new, postfinancial crisis government of President Néstor Kirchner, a multibillion-dollar investment plan in nuclear energy was announced. The investment was justified by reference to the country's spectacular economic recovery and the resulting need for more electricity but also drew on public acceptance that a huge amount of money would be poured into the upkeep of Argentina's two, increasingly venerable, extant plants, a national treasure even if more than 30 years old, and together producing less than 1 percent of Argentina's *energy* needs.

## Brazil

Brazil, the economic powerhouse of South America, operates two venerable nuclear reactors. Angra I was ordered in 1970 from the US supplier Westinghouse (when Brazil was run by a brutal dictatorship supported by the United States) and began operating in 1981. In 1975, Brazil signed with Germany what remains probably the largest single contract in the history of the nuclear industry for the construction of eight 1,300 MW reactors over a 15-year period. However, this time the flirting with the military was a disaster. Due to its ever-increasing debt burden, Brazil failed to keep up with payments, and increasingly blatant interest in nuclear weapons by the Brazilian military made the project politically poisonous in Germany.

## United States

In its efforts to revive nuclear ordering in recent years, the US government has relied on government-backed loan guarantees and regulatory commitments to allow nuclear utilities to recover costs from consumers. These conditions allow utilities to borrow the money they need for reactor projects very cheaply. All nuclear utilities in the United States receive significant federal aid. For one thing, they pay US fuel costs, which average about $4 per MW-hour (MWh), a figure that many see as artificially low because the US government assumes responsibility for disposal of spent fuel in return for a flat fee of $1 per MWh. This arbitrary price was set more than two decades ago and is not based on actual experience.

Otherwise, in the United States, nuclear energy has faced market pressures very reluctantly. As a general rule, electricity is supplied by private utilities that are supposed to compete on the open market for customers. But one exception is the Tennessee Valley Authority (TVA). This is an electricity company 100 percent owned by the US government in Washington. As a result, it has easy access to loans and no concerns about its credit rating. As the nuclear economist Steve Thomas says: "It is therefore not a coincidence that [TVA] has been at the forefront of efforts to restart nuclear ordering." However, being essentially a political animal, TVA reactors are sensitive to changes in the political wind. The two Watts Bar reactors in Tennessee are an example. Construction started in 1973, but work was continually delayed. The first reactor was finally completed 23 years later, in 1996, at a cost of more than $6 billion, while work on the second unit was effectively halted in 1985 with construction described as 90 percent complete. Ten or so years after work on its sister reactor was finished, work restarted on the plant (in 2007), with a topped-up budget and a new target date for completion of 2013. Completing work on designs such as those for the Watts Bar reactors has greater advantages for operators than starting new projects, since regulatory approval has already been given. Indeed, it is highly unlikely that these grandfatherly designs, now 40 years old, would satisfy today's safety standards—not so much because safety standards are higher but because the rules are taken more seriously today.

## Japan

Japan is another country where grand plans to increase nuclear capacity have not been matched by actual orders. In Japan, huge industrial business conglomerates, or *zaibatsu,* with interests in many industries other than energy, typically license nuclear technology from the United States—that is, from Westinghouse and General Electric. (Arguably, Japan saved those companies from going out of business.) And although it may take up to two decades to get formal approval for a nuclear plant in Japan, once construction starts completion is relatively quick and on schedule. Indeed, for many years Japan was the third-biggest producer of nuclear electricity in the world.

That said, it is a country with a great dislike of all things nuclear, with the horror of Hiroshima and Nagasaki deeply etched into the national psyche. Dr. Shoji Sawada, a theoretical particle physicist at Nagoya University in Japan, recalls that at the time the decision was taken to launch Japan on the path that would eventually lead to Fukushima, most

Japanese scientists considered the technology of nuclear energy either still under development or not established enough to be put to practical use. In fact, he says, "The Japan Scientists Council recommended the Japanese government not use this technology yet, but the government accepted to use enriched uranium to fuel nuclear power stations, and was thus subjected to US government policy."

Indeed, as explained in Myth 3, over the years there have been many nuclear accidents at plants in Japan, accidents that generally are badly mishandled and clumsily covered up, all of which has only added to the public dislike and suspicion of the industry.

## South Korea

South Korea, by contrast, has always been a nuclearphile nation. (Or at least its US-backed rulers are.) It has steadily constructed nuclear plants for over 20 years (no nuclear winter there!), and it now draws over a third of its electricity from nuclear plants (see "The Nuclear Powers" table in Myth 1). Plants actually under construction now are likely to raise the proportion of nuclear generation to about half of all South Korea's electricity. The facilities in South Korea's nuclear nemesis to the north are dedicated to making bombs and produce no electricity at all.

## Vietnam

Vietnam has no nuclear power at the moment. But in 2010, Bloomberg financial news excitedly announced that the United States and Vietnam were holding negotiations on sharing nuclear fuel and civilian nuclear technology, an agreement that would, as the article put it, enable American companies such as General Electric to invest in the Southeast Asian country's atomic industry. Vietnam said it planned to build as many as 13 nuclear power plants with a total capacity of 16,000 MW over the next two decades. What, however, about the risk of a Vietnamese bomb? "If Vietnam chooses, as part of its own self-interest, and exercising its right under the NPT [Non-Proliferation Treaty] to enrichment, that is a decision for them to make," said State Department spokesman Philip J. Crowley. "It's not a decision for the United States to make."

## China

Question: Who is building the next generation of nuclear plants? Answer: Almost all of the new reactors in the world are being built by just one country: China. Most of these orders are based on a vintage French design China purchased in 1980 for its Daya Bay site.

Grandiose targets for the country, such as 40 gigawatts by 2020—no, stop, make that 80!—emanate from the all-powerful Central Committee, aggravating the risks of disaster already there due to poor construction techniques, inadequate training, and indeed earthquakes. But this is China, and Mao himself wrote that the country must embrace high-tech. As with the Three Gorges hydroelectric barrage, nature herself is supposed to yield if required.

As of August 2010, there were 37 new reactor projects under way worldwide; of that number, 23—that is, nearly two-thirds—were in China. Another 6 are in Russia, 5 in South Korea, 2 are, or perhaps we should say were, in Japan, and 1 is in France. For all the talk of new-generation reactors, the reality of the nuclear renaissance is of China and Russia using low-tech home suppliers and relatively old designs, with a systemic disregard for reactor security, environmental impact, and worker safety.

## India

During the 1960s and 1970s, India bought a small number of plants from Western suppliers, but its 1975 nuclear weapons test using material produced in a "research" reactor led Western suppliers to sever all contact with the country. Unabashed, India continued to build plants using the 1960s Canadian design it had earlier purchased. Unfortunately, these plants have a poor record of reliability and often take much longer to build than forecast, so both building and funding costs are correspondingly higher.

Addressing a press conference in 2010, the chairman of India's Atomic Energy Commission, Srikumar Banerjee, looked forward to the "imminent" completion of the Koodankulam nuclear power project (India's most expensive ever) and asked for help to reach a national target of nuclear power as 10 percent of overall energy production by 2030 to 2035. How much cash would he like? About 3 trillion rupees. *Trillion!* Even in US dollars, that is a staggeringly foolish $65 billion.

In 2005, the United States decided that India had been punished enough and that a little weapons proliferation could be tolerated, so it reinstituted technological cooperation in civil nuclear power. Canada, Russia, and France swiftly followed. Since then, India has dangled tempting new schemes in front of foreign suppliers and partners, but none has turned into an actual deal. And so, although the Indian government has recently set a target of 63,000 MW of new nuclear capacity to be in service by 2032, using the very latest fast reactors, heavy-water reactors, and thorium-fueled plants, its goals seem particularly vainglorious and improbable.

### Kazakhstan

In 2009 Kazakhstan became the world's leading uranium producer, accounting on its own a year later for one-third of all world production, a remarkable achievement for a country that remains so poor that it has no electricity grid of its own. The US embassy here noted in 2010 that the government's human rights record was patchy, indeed, that it was "very poor," with torture and killing of opponents common.

### Russia

Russia, like China, has had very ambitious plans to expand nuclear power. In 2008, it planned to commission 26 new nuclear units (about 30 gigawatts) by 2025, but by 2009, this target had already slipped back another five years. In both Russia and China, appearance seems to be more important than reality in nuclear matters. Reactors are begun with great fanfare but are never quite completed. Some of the projects are now more than 30 years in the building!

## Turkey

In Turkey, or earthquake central, where a quake in Istanbul killed some 20,000 people as recently as 1999, developing nuclear power is a national priority. Could it have something to do with suggestions that its neighbor Iran is racing to develop its own nuclear power plants to build bombs? Officially, of course not. Curiously, however, both countries have purchased the same Russian-made reactor, the RusAtom VVER1200, a comparative bargain, although it does lack certain safety features of Western models. Indeed, the Iranian one had to be shut down after just three weeks. "The technical problems were enormous," commented Professor Hayrettin Kiliç, to the French newspaper *Le Figaro,* a paper doubtless sympathetic to the idea that the Turks should have bought French reactors. But Kiliç still probably knows his facts. He says that "the pumping system was faulty and I don't think the Central could survive an earthquake of greater than magnitude 6." Which, in a country crisscrossed by active faults such as Turkey, is simply not good enough. So why would the government buy an inadequate system? "Unfortunately, the government preferred the Russian offer for purely financial reasons, because there was simply no more money there for a better plant," Kiliç explained.

## Israel

The "Jewish entity" has something of a national love affair with nuclear power. Not for the electricity it produces, though; for the bombs. As described in more detail later in this chapter, over the last 30 or so years,

Israel's sole nuclear reactor, the Dimona nuclear plant in the Negev Desert, has secretly spawned over 200 nuclear bombs. When Mordechai Vanunu, a former Israeli nuclear technician, confirmed the existence of Israel's nuclear weapons program with photographs of the secret underground bomb facility that were published in the London *Sunday Times* in 1986, the public learned what the experts and other nuclear powers had long known but had done nothing about. Top-secret US government files, later declassified, show unambiguously that certainly by 1975 the United States was fully aware that Israel had nuclear weapons. But it was not the United States that passed Israel the technology; it was France. And that country made sure it got good money for it.

## France

Speaking of which, it must be said that France, the fourth country in the world to get its hands on the bomb, is still one of the world's most nuclear-enthusiastic countries, even if, as explained in Myth 4, its own nuclear industry is creaking under the cost of maintaining and upgrading reactors—let alone the costs of actually decommissioning them.

France's monster energy corporation, Areva, started out as a purely state entity within the Commissariat à l'Energie Atomique (CEA). Created at the end of World War II, the CEA had for years almost unlimited funds and operated with the secrecy of a body responsible for providing France with its nuclear weapons. Even today, it is often referred to in France as the "state within the state." In 1976, its industrial activities were split off into a separate business called Cogema, which in turn, a quarter of a century later (in 2001), merged with Framatome, the French-American constructor of reactors. This was part of a general effort by the ruling French Socialist party to privatize the CEA by rolling together a disparate number of military and civil nuclear agencies, industrial entities, and companies. Some military reactors and nuclear facilities were kept under the direct control of the state, but France's 58 operating civil reactors were transferred to Electricté de France (EdF), the former monopoly supplier of national electricity and now, like Areva, a semiprivate entity. ("Semi-private" in that about 84 percent of its capital is owned by the state.)

As well as its dowry of energy assets, Areva inherited many debts—or would have except that the then-ruling Socialists simply (and secretively) magicked them off the books, a French tradition that was carried on when the business-friendly UMP government (the French acronym means "Union for the Presidential Majority") of President Nicolas Sarkozy took over in 2007. The problem is that even after one tranche of debt has been erased from Areva's semisecret and semipublic accounts, another group grows, periodically plunging the corporation into crisis.

Today, Areva has been obliged to sell some of its "family silver"—select, profitable parts of its complex and ramshackle empire stretching across the world. The capricious Sarkozy government often fails to provide it with further state bailouts; although it still uses the corporation as a tool of foreign policy and a symbol of national pride, the government itself is struggling to keep the financial markets at bay. One sign of this was Sarkozy's unceremonious firing of "Atomic Annie"—Anne Lauvergeon, the Areva chief, in June 2011.

The only Generation III or III+ reactors to be started yet are Areva's European pressurized water reactor ones (they were renamed "Evolutionary Power Reactors" or non-European export markets), one for the Olkiluoto site in Finland, and one in France itself. The EPR received safety approval from the French authorities in September 2004 and from the Finnish authorities in January 2005. The Olkiluoto order is generally seen as a special case, and it has been suggested that Areva offered a price that might not be sustainable to ensure that its new technology is demonstrated. The buyer, Teollisuuden Voima (TVO), is not a typical electric utility. It is a company owned by the major Finnish industrial and power companies and supplies electricity to its owners on a not-for-profit basis. The plant will have a guaranteed market and therefore will not have to compete in the Nordic electricity market, although if the cost of power is high compared to the market price, the owners will lose money.

France, like nearly all other old-nuclear countries, has essentially stopped building nuclear power plants (barring its ill-starred Normandy EPR) and quietly started importing cheap electricity at times of peak demand. In the winter of 2010–2011, France was importing 9,600 MW of electricity on cold days, the equivalent of about ten nuclear power plants operating flat out.

## Sweden

Vattenfall, the Swedish electricity utility, is state owned and thus, like TVA, has easy access to large sums of money at low interest rates. This helps explain why Sweden has a large and thriving nuclear industry enjoying a competitive advantage over companies partly or wholly owned by private shareholders, such as the two main German utilities, E.ON and RWE. (Similar advantages apply in France, where Coface, the French government's loan guarantee body, offers export credit guarantees and Areva has supported overseas sales with special financing.)

## United Kingdom

And finally, over the North Sea, packed with oil and gas rigs, and indeed these days, wind turbines too, to the United Kingdom. The United Kingdom is a nuclear cherry—a small but particularly tasty-looking market. Both it and United States are seen as pioneers of nuclear power, and, therefore, new orders for nuclear plants in these countries carry additional prestige, particularly because a decade ago, economic experiences with nuclear power in both countries were so bad that it seemed unlikely that orders would be possible. Reviving these markets would be a particular coup.

Successive studies by the British government in 1989, 1995, and 2002 came to the conclusion that in a liberalized electricity market, no one would build nuclear power plants without government subsidies and government guarantees that cap costs. Under Margaret Thatcher's center-right government, fresh from dismantling the native coal industry in the 1980s, subsidies were supposed not to be allowed. Coal came in on barges from wherever it could be extracted cheapest. Electricity, it followed, should be subjected to the same market principles. Ever since then, the UK government has battled to support the nuclear industry without seeming to.

The last element of the government's no-subsidies policy came in February 2010, when the energy minister of the Labour (center-left) government, Ed Miliband, set out the new approach in a report in the London *Times* newspaper:

> The *Neta* system [the UK wholesale market for electricity], in which electricity is traded via contracts between buyers and sellers or power exchanges, does not give sufficient guarantees to developers of wind turbines and nuclear plants. He [Miliband] said that one alternative would be a return to "capacity payments"—in which power station operators would be paid for the electricity they generate and also for capacity made available. The idea of such payments is to give greater certainty to investors in renewable and nuclear energy.

Right on political schedule, a day later, the national economic energy regulator announced:

> The unprecedented combination of the global financial crisis, tough environmental targets, increasing gas import dependency and the closure of ageing power stations has combined to cast reasonable doubt over whether the current energy arrangements will deliver secure and sustainable energy supplies. [ . . . ] There is an increasing consensus that leaving the present system of market arrangements and other incentives unchanged is not an option.

In the United Kingdom, the bottom line for nuclear power, since its inception in 1956 and on up to the planned closure of Sizewell B in 2035, taking into account development and prospective waste management and decommissioning expenditure, shows that for every £1 the consumer paid and will pay, the taxpayer has and will pay an additional £2.

## WHAT HAPPENS AFTER THE PARTY IS OVER?

China and India are big enough to treat themselves to the odd nuclear power station, even if they cannot actually afford to provide clean water or food to their citizens. Yet how on Earth do low-income countries such as Sudan, Nigeria, Ghana, Egypt, Indonesia, the Philippines, and others afford their nuclear medicine? Today, the nuclear lobby and its friends in government triumphantly announce a constant flow of new and massive orders for new power plants, which they are especially pleased to sell to the new, emerging economies of the South. Yet China and India, both of which have already announced spectacular and massive nuclear power plant orders—indeed, nearly all the new-nuclear countries, even the Gulf petrostates apart from their oil and gas sector—are still barely industrialized. Outside the governing cliques, their populations range from poor to very poor. These are lands periodically wracked by food shortages, civil strife and unrest, even civil war and border conflicts with neighboring countries. Most of these countries have a tiny domestic demand for electricity (average electricity consumption per capita can be just one-fiftieth of the OECD countries). In addition, their power transmission and distribution infrastructures are often frail and unreliable. And, of course, they have no previous experience of nuclear power or sophisticated infrastructure and expertise to support it.

Perhaps we can shed some light on the mystery of the developing world's interest in nuclear energy by examining the case of Israel, one of the first clients of the French nuclear industry. Today Israel still does not get much electricity from its nuclear reactor—but it does have over 200 nuclear bombs, produced only semisecretly over the last 30 or so years. In an article for Opednews.com, Joe Parko, a delegate for the Quakers on a "peace mission to Israel and Palestine," recalled his meeting with Vanunu in Jerusalem in March 2005. Vanunu told him:

> I worked from 1976 to 1985 at the Israeli secret underground nuclear weapons production facility at the Dimona nuclear plant in the Negev desert. During

> my time there, I was involved in processing plutonium for 10 nuclear bombs per year. I realized that my country had already processed enough plutonium for 200 nuclear weapons. I became really afraid when we started processing Lithium 6 which is only used for the hydrogen bomb. I felt that I had to prevent a nuclear holocaust in the Middle East so I took 60 pictures of the underground nuclear weapons processing plant some 75 meters under the Dimona plant. I resigned my post and left Israel in 1986. I first went to Australia and then made a connection with the Times in London. After a group of nuclear scientists verified my photos as proving Israeli nuclear weapons production, my story was published in England. A few months later, I was kidnapped by the Israelis in Rome and sent secretly by ship to Israel where I was subjected to a closed military trial without counsel. I was sentenced to 18 years in prison. I spent 12 years in solitary confinement.

Although Vanunu was released from prison in April 2004, he was prohibited from leaving Israel. In 2007, he was again imprisoned for six months for speaking to journalists and foreigners.

The American investigative journalist Seymour Hersh found clear evidence of American connivance and support for Israel's nuclear program, noting in his 1991 book *The Samson Option* that Israel was able to coerce several US administrations into doing its bidding. Hersh describes in depth Israeli access to US intelligence satellite technologies, which he explains as due to "inattention by Washington leaders." He also highlights US policies that ignore the very real presence of the Israeli nuclear arsenal.

An article by the Federation of American Scientists provides more detail on the actual history of the Israeli bomb. It starts by noting that Israel was in love with the bomb right from its earliest days. In 1949, a special unit of the Science Corps of the Israel Defense Force (IDF) made a two-year geological survey of the Negev Desert looking mainly for uranium. Although no significant sources of uranium were found, some small, recoverable amounts were located in phosphate deposits.

The next step was the creation of the Israel Atomic Energy Commission (IAEC) in 1952. Its chairman, Ernst David Bergmann, had long advocated an Israeli bomb as the best way to ensure "that we shall never again be led as lambs to the slaughter." Bergmann was also head of the Ministry of Defense's Research and Infrastructure Division (known by its Hebrew acronym, EMET). Under Bergmann, the line between the civilian nuclear program and military one blurred to the point that the military nuclear science unit, Machon 4, functioned essentially as the chief laboratory for the IAEC. By 1953, Machon 4 had not only perfected a process for extracting the uranium found in the Negev but had also developed a new method of producing heavy water, which provided Israel with a homegrown capability to produce some of the most important nuclear materials.

So far, so bad. But, the US Federation of Scientists explains, for reactor design and construction, Israel needed assistance from one of the big powers. And it sought the assistance of France. Nuclear cooperation between the two nations bloomed throughout the early 1950s, when construction began on France's 40 MW heavy water reactor and a chemical reprocessing plant at Marcoule in southeast France. Throughout the decade, France itself was dreaming of how to obtain and explode its first nuclear weapon—a dream it realized in 1960. But in the autumn of 1956, France was able to provide Israel with an 18 MW research reactor. Alas, the sudden eruption of the Suez Crisis a few weeks later upset the cozy situation dramatically.

Recall that following Egypt's nationalistic claim of the right to close the Suez Canal in July 1956—a decision that, curiously enough, was also related to energy politics, since Egypt was enraged at the withdrawal of US and UK funding for the Aswan Dam hydroelectric project that year—France and Britain plotted with Israel that it should provoke a war with Egypt in order to provide the Europeans with a pretext to send in their troops to occupy and reopen the canal zone.

Israel waged its war successfully against Egypt, but following pressure from the United States, the Soviet Union, and the United Nations, the British and French abandoned their military action. This episode is said to have not only enhanced the Israeli view that it needed an independent nuclear capability to prevent reliance on potentially unreliable allies but also to have led French leaders to feel that they had failed to fulfill commitments made to a partner. French premier Guy Mollet was even quoted as saying privately that France "owed" the bomb to Israel.

On October 3, 1957, France and Israel signed a revised agreement calling for France to build a 24 MW reactor (although the cooling systems and waste facilities were designed to handle three times that power) and, in protocols that were not committed to paper, a chemical reprocessing plant. This complex was constructed in secret, and outside the International Atomic Energy Agency inspection regime, by French and Israeli technicians at Dimona in the Negev Desert, under the leadership of an Israeli army officer, Colonel Manes Pratt of the IDF Ordnance Corps.

Construction was a massive undertaking. At its height, some 1,500 Israelis and French workers were employed in building Dimona. It is said that to maintain secrecy, French customs officials were told that the largest of the reactor components, such as the reactor tank, were part of a desalinization plant bound for Latin America. In addition, after buying as much as four tons of heavy water* from Norway on the condition that it not be transferred to a third country, the French air force secretly flew it to Israel.

Notwithstanding all the secrecy, as early as 1958, according to the *Washington Post,* an American U-2 spy plane flying over Israel had already spotted the unusual construction site near the small town of Dimona. The facility featured a long perimeter fence, building activity, and several roads. Israeli officials initially called the facility a textile plant; they later changed their minds and described it as a "metallurgical research installation."

In May 1960, France began to pressure Israel to make the project public and to submit the site to international inspections, threatening to withhold the reactor fuel unless it did so. French president Charles de

* Heavy water, $D_2O$, is water in which the hydrogen atoms have been replaced with Deuterium. It is useful for making plutonium.

Gaulle was concerned that the inevitable scandal following any revelations about French assistance with the project, especially the reprocessing part of the plant, would have negative repercussions for France's international position, which was already on shaky ground because of its war in Algeria. Indeed, in public at least, twice between 1958 and 1960 de Gaulle ordered a stop to French assistance, but "key nuclear officials ignored him"—a likely story, but one good enough for political purposes.

According to the public account, at a subsequent meeting with the Israeli prime minster, David Ben-Gurion, de Gaulle offered to sell Israel fighter aircraft in exchange for the Israelis stopping work on the plant, and the president came away from the meeting convinced that Israel would accept this more modest military advantage over its neighbors. Over the next few months, Israel worked out a compromise. France would supply the uranium and components already placed on order and would not insist on international inspections. In return, Israel would assure France that it had no intention of making atomic weapons, would not reprocess any plutonium, and would reveal the existence of the reactor, which would be completed without French assistance.

Formally, the deal that was struck was that the French government would no longer support construction work on the Dimona reactor but business contracts with private French companies would remain in force. Business rules were more important than concerns about nuclear bombs, you see. In reality, not much changed—French contractors finished work on the reactor and reprocessing plant, uranium fuel was delivered, and the reactor went critical in 1964.

It is said that US president John F. Kennedy leaned hard on Ben-Gurion and his successor, Levi Eshkol, not to build the bomb, even sending US inspectors into the Dimona reactor starting in 1961. But supposedly the Israeli nuclear leaders outfoxed the inspectors (who always gave advance notice of their visits), even building a second, fake control room. And then there was Kennedy's assassination in 1963 to distract American minds. Thus, by November 1966, Israel had the capability to detonate a nuclear device.

Walworth Barbour, US ambassador to Israel from 1961 to 1973—the bomb program's crucial years—even put a stop to the intelligence collection efforts by military attachés regarding Dimona. When Barbour did authorize forwarding information, as he did in 1966, the messages "seemed to disappear" and were never acted on, even after embassy staff learned that Israel was beginning to put nuclear warheads in missiles.

The actual size and composition of Israel's nuclear stockpile is uncertain and the subject of many—often conflicting—estimates and reports. It is widely reported that Israel had two bombs in 1967 and that Prime Minister Eshkol ordered them armed in Israel's first nuclear alert during the Six-Day War. It is also reported that, fearing defeat in the October 1973 Yom Kippur War, the Israelis assembled 13 20-kiloton atomic bombs.

By the late 1990s, the US intelligence community estimated that Israel possessed between 75 and 130 weapons, based on production estimates. The stockpile certainly included warheads for mobile Jericho-1 and Jericho-2 missiles as well as bombs for Israeli aircraft, and may have included other tactical nuclear weapons of various types. Some published estimates even claimed that Israel might have had as many as 400 nuclear weapons by the late 1990s. Stockpiled plutonium could be used to build additional weapons if so decided. For all of these Israeli nuclear weapons, the "electricity-producing" Dimona nuclear reactor, built by the French with the United States' knowledge, is the source of the plutonium. Israel's nuclear program has given it the opportunity to blow up the world—but only if no one else gets there first. And there's competition for that, hey! Because in that sense, the market has truly been liberalized . . . to the point that even pariah states such as Iran, which cannot even buy spare parts for commercial airliners, are allowed their "civilian" nuclear power.

# MYTH 6

# NUCLEAR ENERGY IS VERY CLEAN

*"Uranium itself is radioactive, though with the major isotope U-238 having a half-life equal to the age of the earth, it is certainly not strongly radioactive. U-235 has a half-life one-sixth of this and emits gamma rays as well as alpha particles. Hence a lump of pure uranium would give off some gamma rays, but less than those from a lump of granite. Its alpha radioactivity in practical terms depends on whether it is as a lump (or in rock as ore), or as a dry powder. In the latter case the alpha radioactivity is a potential, though not major, hazard. It is also toxic chemically, being comparable with lead. Uranium metal is commonly handled with gloves as a sufficient precaution."*

*One of the more practical solutions to nuclear waste. (Press Association)*

"TEN TIGHTLY SEALED DRUMS FILLED WITH RADIOactive waste matter caused a certain amount of panic when the French Atomic Energy Commission announced that they intended to be rid of the drums by dumping them into the Mediterranean Sea," reported the newspapers on December 10, 1960. Panicked, or maybe just very cross, the Nice municipal council threatened an administrative strike; the population of Corsica began to organize a mass demonstration; and hotel owners all along the Riviera raised violent objections. Eventually, the commission removed the drums and stored them in "an unrevealed place," as the papers tactfully put it. Quite possibly they are still there, still emitting their invisible death rays, albeit now under supervision. But then again, they might be lying forgotten somewhere at the bottom of the sea . . . who wants to know?

Instead, it is much better to reassure the public that instead of a 600-mile line of grimy railway wagons filled with coal, just two truckloads of cheap and plentiful uranium (from stable countries, such as Canada or Australia) will run a power station large enough to keep a major city going for a year. That is the crucial comparison to be made, at least according to nuclear energy's supporters. And as to the relatively trivial volumes of leftover waste, there is a very promising plan.

Actually, there are lots of plans—almost as many as there are drums of nuclear waste themselves. But there are still no solutions. The unconscionable truth is that if nuclear plants are potential dirty bombs, nuclear waste is an *actual* time bomb, one ticking away right now. Consider what nuclear waste is: metal containers containing material that generates heat, typically up to about 212°F, that can then cause hydrogen gas to burst the container, scattering radioactive particles far and wide, in the best spirit of Pandora's mythical box. No wonder the costs of keeping this waste bottled up are so high—not that the nuclear companies pay it. Rather it is the general public, both today and in future generations, that bears the cost of dealing with nuclear waste. It is just as well for the industry that those uncountable people, well, do not count.

Now, it is true that nuclear waste is quite small in volume compared to the waste produced by other industries, particularly coal. In fact, nuclear power experts love defending the friendly atom against Black King Coal, pointing to the massive slag heaps and poisoned dumps around coal-fired power plants. But if World Nuclear Association websites allow that an industry-standard 1,000-megawatt (MW) reactor has a parsimonious appetite of about 25 tons per year of enriched uranium, they neglect to mention that this fuel has already created 25,000 tons of waste, upstream and downstream, in the process of being mined and refined. Satellite images of mines reveal a tortured landscape stretching for miles.

What is more, landscapes damaged by coal mining can be transformed into parkland, replete with birds and butterflies, even if it may cost a lot of money. Nuclear waste effectively cannot be rendered harmless, so it presents unique and complex problems for dispersal or storage. Reprocessing, despite its promising name, actually *increases* the amount of radioactive waste, and the elusive goal of nuclear fusion, which could simply burn all the waste up like a solar furnace, remains merely a scientific dream. (Good money to be made researching it, though!)

In reality, there are just three options to cope with the nuclear waste problem:

1. Concentrate and contain.
2. Dilute and disperse.
3. Delay and decay.

For the United States, being the country with the largest supply of nuclear electricity brings with it the honor of having the largest amounts of nuclear waste in need of treatment. To get this radioactive and chemically toxic waste out of sight and out of mind, it is easiest to either secretly drop waste into the sea or bury it deep, deep in the ground. With the former option increasingly frowned on, all of the nuclear-committed countries are looking for what are called "geological and deep mine final repositories." Unfortunately, the costs are bankruptingly high and the technological challenges "daunting," as the experience of the United States perfectly demonstrates.

The story of the Yucca Mountain project, near Las Vegas, encapsulates the myriad problems for very long-term disposal and storage of nuclear waste. The repository was to have been located on federal land adjacent to the Nevada Test Site about 80 miles northwest of the Las Vegas metropolitan area. The proposed site was in the south-central part of Nevada near its border with California on Yucca Mountain, a bleak ridge line. Realizing that the project was likely never to be completed, simply because it had been stalled so long, 16 US power companies filed a lawsuit in the court of appeals in April 2010, arguing that the government should stop charging special fees on their nuclear electricity production to pay for the disposal site. At the time they were jointly paying (or, to be more precise, American consumers were paying via a levy on electricity of about one-tenth of a US cent per kilowatt hour) about $750 million a year into the Yucca fund. Over the 27 years since this levy began, the fund accrued about $24 billion, earning the federal government about $1 billion a year in interest.* In this small way, nuclear energy has been a good investment for the government. The catch is that, meanwhile, the projected costs soared to above $85 billion. But this is the sad reality of nuclear waste dumping in all countries. So, rather than getting closer to solutions, governments turn around one day and say, "The problem suddenly got so bad we couldn't do anything about it. Sorry."

---

* A special committee of inquiry, known as the Blue Ribbon Commission, reported on the situation in 2011 and found that, in fact, federal authorities have taken not only the interest on the money but the principal too and used it for other budget priorities, exactly like ordinary taxes. This demonstrates, in a modest way that Dickens character Mr. Boffin, who made his fortune removing rubbish, would have appreciated, that there is indeed money to be made in everything.

**HOW MUCH NUCLEAR WASTE IS THERE ANYWAY?**

In the United States, the 2,000 tons of radioactive waste a year that were being put aside for the never-to-be-built Yucca repository, up until 2009 when the project was canceled, had built up to a tidy pile of 65,000 tons, provisionally stockpiled near the actual reactors.

Or take the example of the United Kingdom. According to the Great British Green James Lovelock, only one or two trainloads of nuclear waste are sitting around. But the UK government is less sanguine. In 2004, it calculated that there were no less than 3 million cubic yards of British radioactive waste, of which more than half was highly radioactive. About 600,000 cubic yards of the waste was already being stored at temporary facilities at more than 500 locations around the United Kingdom, with the balance inevitably involved in decommissioning existing nuclear power stations.

A similarly expensive story, this time in France, is the tomorrow-never-comes Bure project, an underground rock laboratory in the geological clays of eastern France, penciled in as the eventual resting place for that country's waste. In Sweden, however, the cost-trimming solution is to designate the underground final storage repository at Forsmark (the reactor site where routine radiation checks revealed the Chernobyl cloud to the world), a mere 160 feet or so below ground level, and say it is intended only for short-lived radioactive waste. Finland also buries high-level waste under reactors, a solution that avoids the need to pay to transport it.

Despite decades of research and truly vast quantities of money—often dissimulated as military secret spending—there is no economical way of separating high-level and low-level wastes, transporting them to secure and safe storage, and preventing leakage, loss, or theft. In recent years, the estimated costs of radioactive waste disposal have grown massively in all the most nuclear-committed countries, to the point where they inevitably drive up the costs of electricity from any and all generating sources or systems.

Another typical story, this time from Germany. Back in September 2008, United Press International (UPI) reported that the country was engulfed in a discussion over how to best handle nuclear waste, after it had surfaced that leaks threatened security at a radioactive waste dumping site in the Asse mountain range in Lower Saxony. Over the past decades, adding to the problem, some 3,000 gallons of salty base had been flushing into the site each day, mixing with waste that had leaked.

After a report by Lower Saxony's own environment ministry highlighted the deteriorating state of some 125,000 rusty barrels of nuclear waste, Sigmar Gabriel, the then German environment minister, did not beat around the bush. The dumping site, containing as it did some 100 tons of uranium and 12 tons of plutonium, had "as many holes as Swiss cheese," he said. With engineers predicting that Asse, a former salt mine, could last no more than seven more years before collapse, it was "the most problematic nuclear facility in all of Europe," Gabriel confessed to the German mass-selling daily newspaper *Bild*.

The procedural violations surrounding Asse were so outrageous that state prosecutors decided to launch a criminal investigation into the matter. At an emergency meeting, German ministers agreed that the storage facility would henceforth be run—and the bills paid—by Germany's Federal Office for Radioactive Protection, thus nationalizing the problem, and the costs. They also decided that the site would henceforth be treated according to nuclear laws and not mining laws, meaning that the nuclear waste dumped there in the 1960s and 1970s must be made safe underground for the next 100,000 years. This strategy appeared to require the removal of all the waste, something that would be very costly and quite challenging, the experts advised (doubtless rubbing their hands).

For decades, German energy companies and government agencies had considered Gorleben, also in Lower Saxony, as a potential permanent site for highly radioactive waste. Progress had been stalled because of political differences and public protests surrounding nuclear waste dumping. In the face of continuing and furious public protests, in 2000, the German government opted to stop the research altogether.

Eleven years later, in the wake of Japan's nuclear disaster but also due in large part to the earlier trauma of its "waste" mountain, Germany became the first major world economy to appease the public by agreeing to phase out all nuclear power—a step it had already promised once and then reversed. (Its neighbor to the south, Switzerland, once among the stars of the nuclear firmament, also decided to officially abandon the friendly atom.) Yet even once the reactors are switched off, the waste problem remains.

**AMAZING FACT**

According to a report by Greenpeace, "Nuclear Power: A Hazardous Obstacle to Clean Solutions" (2009), "In the past five years, the estimated costs of radioactive waste disposal grew by $40 billion in the United States and by £27 billion in the United Kingdom, with no guarantees to deliver safe storage at the end."

However, nuclear fuel waste is only the beginning of the problem. Sooner or later entire nuclear reactors have to be shut down and made safe. This process, reassuringly called decommissioning, is conventionally split into three phases. Steve Thomas, the nuclear energy analyst, has summarized the process (looking particularly at the British reactor fleet) in this way:

In Phase I, the fuel is removed and the reactor is secured. The time taken to remove the fuel varies, with plants that refuel off-line (that is, not generating power for the grid) being much quicker. These plants are designed for about a third of the fuel to be replaced in an annual shutdown of a few weeks. Reactors that refuel while online take longer because the refueling machine is designed to constantly replace small proportions of the fuel while the reactor is operating. This requires precision machinery that moves slowly, and removing the entire core can take several years.

Indeed, in a large number of cases, this first step has taken as long as 20 years.

Once the fuel has been removed, the reactor is no longer at risk of what the industry calls a critical event, and the vast majority of radioactivity and all the high-level waste has been removed. However, until this phase has been completed, the plant must essentially be staffed as fully as if it were operating. There is thus a strong economic incentive to complete Phase I as quickly as possible consistent with safety. In technological terms, Phase I is simple: It essentially represents a continuation of what was being carried out while the plant was operating. (By convention, the cost of Phase I does not include dealing with spent fuel.)

In Phase II, the uncontaminated or lightly contaminated structures are demolished and removed, leaving essentially the reactor. Again, this work is routine and requires no special expertise. In economic terms, the incentive is to delay entering this phase as long as possible to minimize the amount that needs to be collected from consumers to pay for it—because the longer the delay, the more interest the existing decommissioning fund will accumulate. The limiting point for this corporate prevarication is when the integrity of the buildings can no longer be assured and there is a risk they might collapse, leading to a release of radioactive material.

In Britain, Phase II is planned to take place 40 years after plant closure. British gas-cooled nuclear reactors are expected to be very expensive to decommission because of their physical bulk, which produces a large amount of waste.

Phase III, the removal of the reactor core, is by far the most expensive and technologically challenging phase, requiring remote robotic handling of materials. As with Phase II, the economic incentive is to delay the work until it is no longer safe to do so. Again, taking the UK example, in Britain, this is expected to result in a delay of 135 years.

At the end of Phase III, ideally, the land can be released for unrestricted use—in other words, the level of radioactivity is no higher than in uncontaminated ground. In practice, this is not always going to be possible; at some "dirty" sites, such as the Dounreay site in Scotland, where one of the first-ever demonstration fast reactors operated, use of the land is expected to be restricted indefinitely because of the high levels of contamination.

Needless to say, very few commercial-size plants that have operated over a full life have been fully decommissioned.

The bottom line is that the decommissioning costs estimated for reactors vary widely, but in some cases are several times more than the plants cost to build originally. In other words, dismantling, removing all hazardous waste, storing it in a safe place, and making the site secure run two, three, or even four times the first costs of building and equipping the reactor. Indeed, that rare example of large-scale (and yet still far from actually completed) decommissioning—Chernobyl—shows the costs are indeed astronomical. Just entombing the plant has cost over $2.5 billion so far, with no end in sight. And so the true costs of decommissioning remain a mystery. The cost of waste disposal in modern facilities is also not well established, especially for intermediate-level and long-lived, low-level waste, due to the lack of experience in building facilities to take such waste.

All this uncertainty is reflected in the way that estimates of nuclear decommissioning costs are quoted. Typically, they are quoted as a percentage of the construction cost (perhaps 25 percent). However, as with so many nuclear finance details, uncertainty about prices is a positive for the nuclear waste business.

Consider the task faced by France, the world's most nuclear enthusiastic (and dependent) nation. Under European Union regulatory pressure, Électricité de France (EdF), the national utility as well as Europe's biggest electricity generator, was obliged to set aside enough money by the end of June 2011 for decommissioning of its 58 nuclear reactors and linked waste facilities. Even a specially crafted EdF rescue plan, as we might call it, by the French government offered the utility only a five-year extension on the deadline. And so, at the end of 2010, according to the utility's annual report, EdF had provisions of over €12 billion for decommissioning, to be held in a dedicated asset portfolio of stock and bond investments.

EdF itself allocated nearly another €2 billion ($2.6 billion in US dollars) in cash for decommissioning.

But when the mighty energy company also asked if it could place half its equity in the French power grid into a fund to pay for the dismantling of nuclear reactors, it aroused suspicions that the firm was in increasingly desperate financial straits, according to press reports in 2011 that quoted anonymous sources, including "a person familiar with the project." One such report by Bloomberg News explained that the financial maneuver was intended to produce on paper about €2 billion ($2.6 billion) to help finance the taking apart of old nuclear reactors. (The source declined to be identified because the plan was confidential, not to mention illegal under market competition rules.) Curiously, the French electricity grid itself is heading toward bankruptcy—it needs an estimated €12 billion upgrade ($17 billion) by 2020, according to its own chief executive, Dominique Maillard. But that is in the real world, which barely impinges on high finance.

Actually, in the real world, another Gallic shortcut that Greenpeace often highlights is the Russian agreement with Areva of France to import large consignments of high-level nuclear waste and dispose of it by loading it all into road containers, which are trucked across hundreds of miles of inner Siberia by "local contractors" before being buried or otherwise put out of mind. Some observers speculate that these contractors might make a bit of extra money by supplying the deadly material to terrorists, but even this risk still pales in comparison to the radiation being released into the environment by the "authorities."

Indeed many reputable energy companies have found it tempting that at least some radiation can be released discreetly back into the environment. For example, according to a paper by the Strategic Studies Institute, the global, collective dose over 100,000 years—due primarily to annual releases to the atmosphere from just one French nuclear reactor at La Hague—of the low-level but long-lived emitters krypton-85 (which has a half-life of 11 years), carbon-14 (5,700 years), and iodine-129 (16 million years) has been calculated at 3,600 man sievert. Continuing discharges at this level for the remaining years of La Hague's operation could theoretically cause some 3,000 additional cancer deaths . . . but no one would ever know.

Any complete survey of the nuclear waste issue should also include other options that have been used in the past and doubtless continue to be used today. The shipment of low-level waste to African countries, perhaps hidden among household, hospital, and industrial waste, is one quick fix that has been used; the dumping of nuclear waste in the sea—and simple dumping nuclear scraps and leftovers on wasteland along roadways or as fill, perhaps for shopping centers, perhaps for recreation centers—is another.

In the 1950s, one by-product of the nuclear industry, called depleted uranium (DU), actually became very fashionable. It was made into key fobs, dice, even golf clubs. However, as it is, well, radioactive, it fell out of fashion.

So nowadays only one truly economic way to deal with this component of nuclear waste is left, and that is to use it as weapons. This is by no means a marginal activity, undertaken only by bearded Islamists. Rather, it is a mainstream business option of the western nuclear powers. It is the only kind of nuclear waste disposal that actually improves the economics of the nuclear industry—a single 155 mm antitank shell tipped with depleted uranium can fetch about $6,500.

The original blueprint for DU weaponry is a 1943 Manhattan Project memo to General Leslie Groves that recommended development of radioactive materials as poison gas weapons: dirty bombs, dirty missiles, and dirty bullets. In recent years, depleted uranium has been used as sheaths or nose cones on shells and missiles. Size 105 mm, 155 mm, and 203 mm artillery shells are the ones typically modified, to become high-energy antitank (HEAT) ordnance, as are rocket-propelled grenade shoulder-launched missiles, such as Milan-type or Eryx-type missiles. The United States has used the weapons in Yugoslavia, Afghanistan, and twice in Iraq since 1991, describing them as "conventional." The United States has also supplied Israel with DU weapons; Israel used them on inhabitants of the crowded cities of Gaza who are guilty of periodically launching homemade rockets at their neighbors.

On the battlefield, as well as for years afterward, DU has three effects on living systems: It is a heavy metal chemical poison, it is a radioactive poison, and last but not least, it has a particulate effect due to the very tiny size of the particles (0.1 microns and smaller) that causes additional diseases in humans. Like the chemical and biological weapons banned after World War I, DU weapons are totally indiscriminate and can end up killing those who originally fire them.

Since the effects are so awful, why is DU used? The answer is simple: DU is used because it is a cheap and abundant by-product of the nuclear power industry.

Today the world annual DU output is probably well above 35,000 tons, meaning that at least 30,000 tons a year are accumulating, roughly 60 percent in Europe and the United States. Over ten years, a massive stockpile is generated, and storing DU costs money. It is slightly radioactive, chemically poisonous, and can catch fire, and rogue nations supposedly want to get their hands on it to make dirty bombs. So, much better to put it to the service of the democracies!

As Jalal Ghazi, a journalist who monitors and translates Arab media for New America Media, pointed out recently:

> The U.S. and British militaries used more than 1,700 tons of depleted uranium in Iraq in the 2003 invasion (Jane's Defence News, 4/2/04)—on top of 320 tons used in the 1991 Gulf War (Inter Press Service, 3/25/03). Literally every local person I've ever spoken with in Iraq during my nine months of reporting there knows someone who either suffers from or has died of cancer.

Ghazi went on to report that in Fallujah, which bore the brunt of two massive US military operations in 2004, as many as 25 percent of newborn infants have serious physical abnormalities. Cancer rates in Babil, an area south of Baghdad, have risen from 500 cases in 2004 to more than 9,000 in 2009. Dr. Jawad al-Ali, the director of the Oncology Center in Basra, told Al Jazeera English (10/12/09) that there were 1,885 cases of cancer in all of 2005; between 1,250 and 1,500 patients visit his center every month now.

This sad report may help explain the lingering question of the mysterious Gulf War syndrome that affected so many American and Allied soldiers after they returned from service in the Middle East. As to all this perplexity, in a statement, the Pentagon merely said: "No studies to date have indicated environmental issues resulting in specific health issues. Unexploded ordnance, including improvised explosive devices, are a recognized hazard."

DU is the Trojan horse of nuclear war—and once allowed into the citadel, it is one that keeps killing. There is no way to clean it up and no way to turn it off, because it continues to decay into other radioactive isotopes in 20 or so sinister steps.

To be sure, the well-paid and highly sincere nuclear lobbyists will steer far away from lauding this particular quick fix for the nuclear waste crisis. They prefer disposal concepts bordering on technological utopia, including jettisoning waste into space and "burning" it in high-neutron-density experimental reactors.

For example, some US researchers now think they know how to convert nuclear waste into far safer elements with a hybrid reactor. Physicists at the University of Texas at Austin have designed a new system that, when fully developed, would use fusion to eliminate most of the transuranic waste produced by nuclear power plants. "We have created a way to use fusion to relatively inexpensively destroy the waste from nuclear fission," promises Mike Kotschenreuther, senior research scientist with the Institute for Fusion Studies. Only, the scientists have not quite gotten there yet. More money please!

Actually, the research that is always cited as making this solution feasible is far from groundbreaking. The concept of fast neutron "burn-up" reactors dates from the early 1960s. As the pronuclear World Nuclear Association says on its website:

> Fast neutron reactors are a technological step beyond conventional power reactors. They offer the prospect of vastly more efficient use of uranium resources and the ability to burn actinides which are otherwise the long-lived component

> of high-level nuclear wastes. Some 390 reactor-years experience has been gained in operating them.
>
> About 20 Fast Neutron Reactors (FNR) have already been operating, some since the 1950s, and some are supplying electricity commercially. About 390 reactor-years of operating experience have been accumulated. Fast reactors more deliberately use the uranium-238 as well as the fissile U-235 isotope used in most reactors. If they are designed to produce more plutonium than they consume, they are called Fast Breeder Reactors (FBR). But many designs are net consumers of fissile material including plutonium.
>
> Fast neutron reactors also can burn long-lived actinides which are recovered from used fuel out of ordinary reactors.

The website lists 18 fast neutron reactors in eight countries, the oldest of which date back to 1951 (in the United States), which looks like an impressive statistic, but, actually, of these 18, only 5 are still in operation. One basic reason for the abandonment of this research is the almost open-ended cost of burn-up reactors, combined with their very unimpressive performance.

Miraculously for the industry, however, the huge costs of these high-tech reactors, which are closely related to the similarly unworkable and costly fast breeder reactors, never impinge on or migrate into the list of costs and charges for civil nuclear power plants. Consider the international "fusion machine" project at Cadarache in southern France. This project, already subject to periodic revisions of its final costs and completion date for operating an experimental and small-scale fusion reactor, is also presented as being a potential burn-up reactor for nuclear wastes—even if it cannot be made to work as a fusion reactor.

Of course, there are still some diehard nuclear enthusiasts for whom even the nuclear waste issue is not an insurmountable problem. One such person is the green energy guru George Monbiot of the *Guardian* newspaper. Ironically, at one time, of all the possible concerns about nuclear, it was the waste disposal question that preoccupied him the most. No

### THE UN PLAN

According to the International Atomic Energy Authority, the internationally defined options for decommissioning include:

- *Immediate dismantling* (or early site release in the United States). This option allows for the facility to be removed from regulatory control relatively soon after shutdown or termination of regulated activities. Usually the final dismantling or decontamination activities begin within a few months or years, depending on the facility. Following removal from regulatory control, the site is available for reuse.
- *Safe enclosure* (or *Safestor* in the United States). This option postpones the final removal of regulatory controls for a longer period, usually in the order of 40 to 60 years. The facility is placed into a safe storage configuration until the eventual dismantling and decontamination activities occur.
- *Entombment or "sarcophagus containment."* This option entails placing the facility into a condition that will allow on-site radioactive material to remain there without ever being totally removed. This option usually involves reducing the size of the area where the radioactive material is located and then encasing the facility in a long-lived structure, such as concrete, that will last for a period of time to ensure the remaining radioactivity is no longer of concern.

longer! In one of his keynote articles on nuclear power, he put the issue in Freudian terms:

> The most fundamental environmental principle, taught to every child before their third birthday, is that you don't make a new mess until you have cleared up the old one. It seems astonishing to me that we could contemplate building a new generation of nuclear power stations when we still have no idea where the waste from existing nukes will be buried.

Now, as part of his maturing process, it seems the "big block" has been removed. Regarding the "mess" created by nuclear waste, Monbiot continues:

"It's true that my position has changed. As the far more terrifying effects of climate change have become clearer, nuclear power, by comparison, has come to seem less threatening."

Several things changed his view, but his worries about the risks were finally assuaged after reading the technical report by the Finnish radioactive waste authority Posiva in particular. (Evidently, of course, a neutral assessment.) "This seems to me to be a convincing demonstration that the long-term storage of nuclear waste could, in principle, be carried out safely," Monbiot says.

So what is this Finnish nuclear solution upon which hang the hopes of the industry? Essentially it is to bury the waste about 500 yards down in an area of particularly stable rock, in specially constructed copper containers, and surround the lot with a special layer of clay.

In a generally enthusiastic account of the enterprise titled "Finland's Nuclear Waste Solution: Scandinavians Are Leading the World in the Disposal of Spent Nuclear Fuel," journalist Sandra Upson describes the $4.5 billion project:

> What the company has bet on is a nested system of what it calls engineered barriers, which are enveloped by the natural barrier of gneiss bedrock. The first engineered container for the radioactive refuse is the copper burial cask, within which sits an iron insert. Each canister will then be buried in specially dug holes in the underground tunnel network and surrounded by a special clay—the second engineered barrier—through which water can slowly diffuse, but not flow. A century from now, after Finland's last planned reactor has long been closed and its fuel has cooled, the tunnel's empty spaces will be filled back up with rubble and clay, the final safeguard. A concrete slab will cover the entrance and, the designers hope, deter future adventurers.

So, if you are not convinced by anything else, the big slab should reassure you! It's *Tomb Raiders of the Lost Ark.* The plan does not really envisage any protection against earthquakes—although an iron lattice is added to strengthen the copper containers—but rather assumes there will be none. (In fact, earthquakes do occur in Scandinavia, a fact that brings the possibility of the casks being breached well within the kinds of risk parameters that nuclear power stations themselves have repeatedly failed to reassure on.)

As with all deep-repository schemes, the big problem with the Finnish solution is the risk of groundwater getting to the containers. Once there, even a microscopic defect is enough for water to infiltrate and start its mischievous work. (Particularly as the company envisages that some of its special copper casks will likely be damaged in the process of rolling them deep underground into their nuclear crypt.) These chemical and bacterial processes, perhaps aided by residual heat from the spent fuel inside, could conceivably make nonsense of the grand "hundreds of thousands of years" of security promised by the site developers.

As Upson herself notes in what she calls "the nightmare scenario": "[W]ater would somehow manage to reach the canisters, carrying with it bacteria that burrow through the clay and erode the metal containers. The fuel rods would become exposed to the clay, and the water would carry harmful radionuclides from the fuel back to the surface."

Or, indeed, back to the North Sea. The proposed deep-burial site is not only situated in an area with several earthquake faults, not only situated in rock running with brackish water, but is part of a complex circulatory flow that sees rainfall traveling in a perpetual circle down from the surface of the Finnish nuclear island and then back up later to the sea. If the Finns have gotten their calculations wrong, it will not be just the few inhabitants of Olkiluoto Island, where "the forest is king, elk and deer graze near sun-dappled rivers and shimmering streams" and humans "search out blueberries and chanterelle mushrooms," who need to worry.

# MYTH 7

# NUCLEAR RADIATION IS HARMLESS

*"We are surrounded by naturally occurring radiation. Only 0.005 percent of the average American's yearly radiation dose comes from nuclear power; 100 times less than we get from coal, 200 times less than a cross-country flight, and about the same as eating one banana per year." Or so, no doubt, would say Sister Mary Helene van Horst, science instructor, pictured here, top left, teaching students all about radiation and monitoring at a college in Iowa deep in the Corn Belt of the United States, in 1960.* Note the carefully locked box with the dangerous samples of radioactive substances in the background.*

* It might seem incongruous that a nun should teach radiation class, but in at least one way it makes very good sense: The people most at risk from nuclear radiation are not adults but rather unborn children whose mothers have come into contact with radioactive particles. And nuns do not have children.

*Montage based on images from the US National Archives. (US National Archives)*

**ON TODAY'S PRONUCLEAR WEBSITES, JUST AS IN BY-**gone civil defense classes, lesson one always offers a reassuring answer to the question: "How dangerous is radiation anyway?" But here is a worry that simply refuses to go away. And the answer remains crucial to the economic viability of nuclear power. Because even when nuclear reactors do not blow up, leak, or melt down, they create massive quantities of radiation. Each new, state-of-the-art reactor produces over 75 pounds (35 kilograms) of plutonium and over 30 tons of other high-level radioactive waste each year inside its fuel rods and reactor assemblies. These materials will need to be extracted, reprocessed into reactor fuel, or stored in plutonium repositories (for the plutonium) and high-level waste repositories (for the rest).

What this means is as simple as it is devastating: Each reactor contains what in industry jargon is called a "radiological inventory" equivalent to 150 or 250 times the total radiation release of the 1945 Hiroshima atomic bomb.

For some, such as Helen Caldicott, an undisguised opponent of the industry:

> There is no way to put it on earth that's safe. As it leaks into the water over time, it will bioconcentrate in the food chains, in the breast milk, in the fetuses, that are thousands of times more radiosensitive than adults. One X-ray to the pregnant abdomen doubles the incidence of leukemia in the child. And over time, nuclear waste will induce epidemics of cancer, leukemia and genetic disease, and random compulsory genetic engineering. And we're not the only species with genes, of course. It's plants and animals. So, this is an absolute catastrophe, the likes of which the world has never seen before.

Yet the industry itself has a very different view, spread around liberally by expert consultants and media advisors. Fear of nuclear radiation is irrational and superstitious, they maintain.

After all, as every nuclear physicist will tell you, radiation is entirely *natural.* Radiation is all around us, in the rocks, in the sunlight, in the soil—why, there is even a bit in bananas! The second thing they will earnestly explain is that our friendly Sun produces huge quantities of radiation because, after all, it is *also* a nuclear reactor, busily fusing atoms together with heat energy as a by-product. (Although they probably will not mention that the natural radiation from the Sun would still kill all life on Earth were it not for the happy coincidence that our planet happens to have a large molten iron core, which generates a protective magnetic field that shields us from the radiation.)

The second reassuring thing they will tell you is that although, yes, one spooky fact about radiation is that some of it can go through brick walls and even serious metal cladding, the rest is feeble stuff. A piece of paper—or, more to the point, human skin—is enough to block it. And we should rest assured that all of the dangerous radiation produced by nuclear plants will be locked away securely in steel drums deep underground.

Much of this is true. And so the final thing the experts say is that fear of nuclear power is out of all proportion to the actual risks, born of ignorance and superstition.

Recall again the words of Green Guru and Respected Scientist James Lovelock on the lessons he drew from Chernobyl, written in May 2004:

> Opposition to nuclear energy is based on irrational fear fed by Hollywood-style fiction, the Green lobbies and the media. These fears are unjustified, and nuclear energy from its start in 1952 has proved to be the safest of all energy sources. We must stop fretting over the minute statistical risks of cancer from chemicals or radiation. Nearly one third of us will die of cancer anyway, mainly because we breathe air laden with that all-pervasive carcinogen, oxygen. If we fail to concentrate our minds on the real danger, which is global warming, we

> may die even sooner, as did more than 20,000 unfortunates from overheating in Europe last summer.

Given all that, why, today, numerous official reassurances later, are so many people still terrified of atomic radiation? Perhaps because it is invisible, which makes it all the more threatening, and because of its potential to cause cancer and genetic deformities. Other cancer-causing agents such as food or smoke seem innocuous by comparison, more tangible, more familiar. Nuclear accidents such as those at Three Mile Island, Chernobyl, and Fukushima still provoke scary headlines . . . because they are events from a world most people know little about.

Not that the media is especially critical. Indeed, on such complicated issues, they are generally quite happy to be led by the authorities. Literally. So, for example, in the radioactive shadow of Fukushima, the British Department of Energy (a body naive folk might have supposed to be concerned with safety issues) quickly worked out a strategy with its "industry partners" on how to swamp the media with pronuclear feature articles. (A fact that only emerged thanks to UK Freedom of Information laws.) Soon, sprouting like so many mushrooms, reassuring authorities appeared on the op-ed pages of the British papers conveying the message that "nuclear remains one of our safest and cleanest sources of electricity." After one such dollop of hubris in *The Times* (London), the newspaper's science editor weighed in, saying firmly that nuclear power is "considerably safer than some of the alternatives" and that the really important thing was for governments to upgrade to new nuclear reactors. In the *Daily Mail,* among its diet of celebrity gossip and true crime stories, appeared Sir Max Hastings, a man better known for his war reporting, patiently repeating the nuclear myths one by one and emphasizing that Fukushima had so far killed no one—not one! A few small explosions, yes, but no *British* casualties. (We paraphrase.) "Yes, nuclear power plants are dangerous. But for Britain, the alternative is to start hoarding candles," boomed Hastings, explaining that nuclear is a key component of a "credible energy future." "In a world dominated by

### HOW MANY PEOPLE WOULD DIE IF THERE WAS A NUCLEAR ACCIDENT IN THE UNITED STATES TODAY?

During the late 1980s, the Nuclear Energy Agency (NEA), a specialized agency within the Organization for Economic Cooperation and Development (OECD), an intergovernmental organization of industrialized countries, based in Paris, carried out perhaps the most comprehensive study ever of the consequences of a nuclear accident in the United States.

The study, known as CRAC-2, or in full "Calculation of Reactor Accident Consequences 2" (essentially, a computer program), was carried out by the Nuclear Regulatory Commission. It estimates the impacts from a meltdown at each nuclear plant in the United States in categories of "peak early fatalities," "peak early injuries," "peak cancer deaths," and "costs" and prices it in billions of 1980 dollars. (Price inflation means these figures may be multiplied many times now.) "Peak" refers to the highest calculated value—not a "worst-case scenario," which could be much worse. For the reactor complex known as Indian Point, north of New York City, for example, the projection is that a meltdown would cause 50,000 "peak early fatalities," 141,000 "peak early injuries," 13,000 "peak cancer deaths," and $314 billion in property damage—perhaps $1 trillion in today's money. For the Salem 2 nuclear plant in New Jersey, the study projects 100,000 "peak early fatalities," 70,000 "peak early injuries," 40,000 "peak cancer deaths," and $155 billion in property damage. The study found similarly staggering numbers across the country . . . yet the statistics scarcely ever made the papers, presumably because the threat from organized crime, terrorism, et cetera is so much more pressing.

headlines and TV images, we are extraordinarily bad at measuring risk sensibly before making vital decisions." Finally, Hastings pointed out that "while thousands of Japanese people have died in the tsunami, there is still no evidence that anybody has been, or will be, killed by fallout" and assured his readers that "accidents in the oil and coal industries have killed and continue to kill far more people than nuclear power ever has."

To complete the triangulation, scientific experts such as Hidehiko Nishiyama, of the Nuclear and Industrial Safety Agency in Japan, explained to the world's press that "[o]cean currents will disperse radiation particles and so it will be very diluted by the time it gets consumed by fish and seaweed." (One of the stranger things about the radiation debate is that views seem to get more polarized over the years and the truth gets harder and harder for anyone to find, despite the reasonable assumption that as a technology matures and becomes more common, we all have *more* information, not less.)

At least one thing everyone does agree on is that the invisible specter of radiation and the cost of dealing with all the radioactive waste it generates are a ball and chain on the industry. In fact, the industry can survive only by giving an optimistic assessment of the dangers of low-level radiation. Farmers and fishing communities that find a nuclear plant springing up in their midst, pumping slightly irradiated water into the seas or rivers and releasing slightly irradiated steam into the atmosphere, all need constant reassurance that the amounts are inconsequential compared to "normal" background radiation.

Added to that, nuclear reactors employ surprisingly large numbers of people, and it takes money to persuade someone to work in a hazardous environment. It would be so much easier (read: profitable) to run a nuclear reactor if radiation was really not that dangerous after all.

First of all, it would lower staff costs—but maybe not very much. Because it turns out that if our collective image of a nuclear power station consists of highly trained technicians in white coats, the reality is teams of casual workers being recruited from building sites and farms for short contract work. Take, for example, the world's third-largest generator of nuclear power, the supposedly high-tech Japan. According to Hidehiko Nishiyama's Nuclear and Industrial Safety Agency, Japan's nuclear plants have for many decades been staffed almost entirely by casual workers.

In fact, contrary to the image, the people who *really* keep nuclear power plants running are not the physics graduates in the high-tech control rooms but the Homer Simpsons paid by the hour who otherwise might work in agriculture, on building sites, or on roads. Their jobs consist of cleaning radioactive dirt off the sides of reactors' dry wells or mopping out spent-fuel ponds with mops and rags—all the while receiving higher rates of pay (in Japan wages start at $350 a day and may go up to $1,000 for just two hours) against perceived dangers in the work. While there is an elite of nuclear industry employees, who are highly trained and carefully monitored for radiation, these workers are nonunionized and frequently obliged to falsify their radiation records.

In March 2010, of the approximately 83,000 workers in the 18 commercial nuclear installations in Japan, 88 percent were contract workers. At the Fukushima complex, the proportion of casual staff was even higher.

Tetsuen Nakajima, the chief priest of a 1,200-year-old Buddhist temple, once attempted to set up a union for the huge pool of casual nuclear workers in Japan. Among the union's strikingly modest ambitions were that workers should not be made to lie to government inspectors about safety procedures and that workers' radiation records should not be routinely falsified to enable them to work longer. But his embryonic union was dissolved when unknown thugs appeared, threatening violence against anyone who joined. As Nakajima put it in comments to a newspaper, "They were not allowed to speak up. Once you enter a nuclear plant, everything is a secret."

And one of the best-guarded secrets, without any doubt, is how much radiation nuclear plants are releasing into the environment. That is why, even after Chernobyl exploded, Soviet authorities made no public announcements in the crucial first days, but only after instruments at the Forsmark nuclear plant in Sweden, over 600 miles away, detected the cloud; it is also why, when Fukushima started to melt down, spreading radioactive particles over Japan and over the Pacific to the United States, it was the United States, not the Japanese authorities, that raised the alarm.

Then, as the Fukushima plant continued to belch radioactive particles into the air, authorities moved quickly to ban milk in Fukushima itself, the epicenter, so to speak, of the contamination, followed by spinach from neighboring Ibaraki to the south; canola from Gunma in the west, and finally chrysanthemum greens from Chiba to the south.

The day after the tsunami set off the disaster at the nuclear plant, residents of the nearby town of Namie were ordered to assemble, ready to evacuate. Officials told families to relocate to the north, which for three nights is exactly what they did, sheltering in a district called Tsushima. As hydrogen explosions at four of the reactors spewed clouds of radioactive particles into the air, their children played outside, and residents relied on temporary facilities, including outdoor washing facilities. Meanwhile, the government in Tokyo examined computer projections that showed the radiation cloud extending in a 40-mile tongue straight toward where the residents had been evacuated to. But the government kept this information to itself.

Said the mayor of Namie some months later, when the truth finally emerged: "From the 12th to the 15th we were in a location with one of the highest levels of radiation. We are extremely worried about internal exposure to radiation." In this case, the nuclear secrecy, he said, was tantamount to "murder."

The principal hazards released into the environment at Fukushima were iodine-131 and cesium-37. Iodine is short-lived, and the effects can be mitigated to some extent if the public is promptly warned to avoid certain products and to start taking iodine tablets. Mind you, the precautions would need to be widespread. For example, a few days after a nuclear weapons test in China in the 1990s, scientists detected radioactive iodine in deer in the United States.

But we cannot do better than to look at the case of Chernobyl to see the broad array of interpretations about radiation hazard. For truly, Chernobyl is the front line between popular opinion, which says its meltdown was an unparalleled disaster, and expert opinion, which says it was a mere hiccup in energy production.

As noted already, three days after Chernobyl, the Soviet authorities had made no announcements whatsoever. The government continued to suppress news of the incident, even as it began to forcibly evacuate people living nearest to the plant. Farmers continued to distribute milk and other products that now contained within them "the seeds of death," as one commentator rather colorfully put it.

Later on, though, an exclusion zone was created, which became a massive laboratory for measuring radiation effects. (Not all crops are equally affected by radiation. Those with bigger leaves, such as lettuces, spinach, and grass, generally absorb most, while crops such as potatoes, rice, and corn absorb less. Fruits such as apples and oranges also collect less radiation. Clay soils tend to trap radiation in the soil whereas sandy soils allow it to pass more quickly into plants.) Trees eventually trapped much of the radioactivity, as did mushrooms, some of which are said to be 1,000 times more radioactive than is either normal or safe to ingest. Ornithologists noted that swallows, which migrate and hence are in the region only a part of the year, exhibited many tumors and genetic defects. The conifers seemed to be particularly sensitive to the radiation and were steadily being replaced by silver birch. Indeed, over the years, many studies have been undertaken, but the conclusions are often contradictory and in opposition to one another. However, one plausible theory is that generally the plants and animals that are the most ancient in origin—that is, with the simplest and hence most resilient genetic code—survive radiation best. An ancient line of wild horses has been reintroduced to the area, adding a touch of exoticism.

These days, the fear of the invisible radiation has dropped considerably in the region around Chernobyl. Hardy peasants have even reclaimed some of their homesteads. A panel of experts commissioned by the World Health Organization (WHO) to look into the health consequences concluded that the problems were much exaggerated and trivial.

Nowadays, tourists regularly tour the 775-square-mile national park that is the exclusion zone, admiring the great sarcophagus itself and the abandoned apartments and ruined fairgrounds of the nearest towns. But if they are offered chicken and mushroom pie with side servings of locally produced spinach, they should decline and ask for apple tart instead. And in any case, they must depart the zone through military-style checkpoints by 8 p.m., leaving the wolves and wild boar to roam the ruined streets in peace.

The Chernobyl meltdown created an especially large radiation cloud—because the reactor itself was very large. Even as suicide helicopter pilots valiantly managed to put out the fire in the reactor core and then entomb the broken reactor in a coffin of sand, boron, and concrete, a cloud of radiation quickly covered most of western Europe, stretching from the beaches of southern Spain to the highlands of Scotland.

In due course, 5 million people would be exposed to radiation in Belarus, Ukraine, and Russia alone. Of these, some 250,000 workers (including both army conscripts and volunteers (note that army conscripts are *not* volunteers) received particularly high doses. The people who took part in cleanup operations around the reactor and in towns and villages considerably farther away became known as "the Liquidators."

Tracking radiation is itself a tricky task. Although after a nuclear incident a circle will be drawn on the map around the affected plant, 5, 10, or perhaps even 20 miles distant, and although large numbers of people within that circle will be forcibly moved—a quarter of a million residents in the case of Chernobyl—the actual radiation follows no such pattern. It follows the winds and the tides, shooting off in a concentrated plume for hundreds or even thousands of miles before deciding to settle. Of course, no one tells you that. After all, no one has any solutions.

Once the particles land, perhaps brought down by rainfall, their journey is by no means ended. They continue to travel along food chains and biochemical pathways for decades to come.

Does it matter, though? WHO experts in the West say no, but others nearer the actual events beg to differ. When Alexander Veremchuk, a doctor at a hospital specializing in radiation diseases in Kiev, surveyed the situation in the 30 main hospitals of his region 20 years after the accident, he found that "up to 30% of people who were in highly radiated areas have physical disorders, including heart and blood diseases, cancers and respiratory diseases. Nearly one in three of all the newborn babies have deformities, mostly internal."

Other unfortunate laboratories for testing the effects of radiation have been the Pacific islands where nuclear bombs have been exploded. Not, of course, that all Pacific islands have been used for testing nuclear bombs, although there are more than you might imagine. The 2,000-plus tests worldwide are equivalent to one test every nine days for the last 50 years. Indeed, between July 16, 1945, and September 23, 1992, the United States alone conducted (by official count) 1,054 nuclear tests (not counting the two nuclear bombs dropped on Japan).

The peak years were at the height of the Cold War, when major changes in weapon design occurred. During this time, tests were grand operations, involving huge numbers of people. (By the time of the last test series—Dominic I—some 28,000 military and civilian personnel were involved.) Explosions were carried out in all environments—aboveground, underground, and underwater. On top of towers, on barges, from balloons, in tunnels, and fired by rockets hundreds of miles up into the atmosphere. It is estimated that over 9,000 pounds of plutonium have been discharged into the atmosphere as a result. Which is where you least want it to be, as plutonium is primarily dangerous when inhaled.*

The Marshall Islands in general, and Bikini Atoll in particular, are famous for their nuclear tests. The islands were directly administered by the United States for decades, until they nominally became a sovereign state in 1986 with the signing of the Compact of Free Association with the United States.

The first US air drop of a thermonuclear weapon was made at Bikini Atoll in May 1956, in a test known as Cherokee. This test was mainly intended to impress the Soviet Union. The final Flathead test of June 1956 at Bikini Atoll lagoon is of more interest to the nuclear industry. This one was a test of the easiest doomsday machines to construct—a cobalt

* Research on the toxicity of inhaled plutonium draws heavily on the "lessons" of Hiroshima, but also on spectacularly cruel experiments on beagles in laboratories. It turns out that when obliged to inhale small amounts of radioactive plutonium the dogs suffer a wide range of illnesses such as lung, liver, and skeleton tumors and radiation pneumonitis.

bomb cluster. Each cobalt bomb is an ordinary atomic bomb encased in a jacket of cobalt. When the bomb explodes, it spreads a huge amount of radiation.

Here, as in Chernobyl's exclusion zone, nature proved resilient. Half a century later, Bikini Atoll looks again like a tropical paradise, a jewel in the ocean where palm trees framing a blue-green lagoon sway in the breeze. But also like Chernobyl, the effect on local human populations has been longer lasting and crueler. During the tests, many people living on and near Bikini and Eniwetok atolls were forcibly evacuated to other islets. Since these locations were isolated and barren, the natives were forced to exchange their traditional "canoe" culture of fishing and gathering fruit for a new US one of food stamps and processed-food handouts. For those who were evacuated, it seemed a great injustice, scarcely made any better by the fact that 239 people were left behind, presumably by accident. These islanders then faced the blast radiation of the first *hydrogen* bomb as part of the 1954 Bravo! test, the United States' biggest bang ever. Some islanders still remember seeing the "second sun"—an intense fireball 1,000 times more powerful than Hiroshima, and the 20-mile-high mushroom cloud.

It was followed by hurricane-force winds that stripped the branches and coconuts from the trees. To the delight of the planners, a small fleet of empty ships, including the USS *Saratoga* and the *Nagato,* the flagship of the Japanese admiral Isoroku Yamamoto (the man who planned the attack on Pearl Harbor), were engulfed in the explosion and plunged to the bottom of the lagoon. And a plume of radioactive fallout spread quickly toward Rongelap, an unevacuated island nearby. A fine powder fell like snow (not that the islanders had ever seen snow), and anyone whom it touched became nauseous, and their skin turned red and peeled off in layers. In the years that followed, just as in the Ukraine, humble people would measure the health effects in numbers of cancer deaths and birth defects per family.

As for normality returning to the exclusion zones, whether it is around Chernobyl or Fukushima, well, do not hold your breath. The Bikini Islanders who were evacuated were promised it would be for a short period only; indeed, the Pentagon optimistically reported that after the tests, "All of the test islands have been swept clean and Elugelab in particular is completely gone, nothing there but water and what appears to be a deep crater." But when the first 150 people were returned to Bikini in 1967, they had to be evacuated again quickly after it was discovered that radioactive cesium-137 still contaminated everything. Since cesium-137 has a half-life of 20 years, for the radioactivity to drop to acceptable levels will take a good few generations. Restoration of the Bikini Atoll was attempted once more in 1969, but again the vegetation was found to be contaminated and highly toxic.

It seems that tourists can now visit the atoll (but are warned against swallowing anything). Dive enthusiasts can even swim around the lagoon looking at the sunken US and Japanese aircraft carriers, warships, and submarines. But for the islanders, such novelties have worn off. On Ejit Island, where most of the Bikini Islanders ended up, barefoot children play outside shanty homes with nothing else to do.

The United States put little effort into checking on the health consequences of the various nuclear bombs on the Pacific islanders. The health consequences on the world's population are conveniently impossible to calculate, although a study by the US National Cancer Institute in 1997 found that, given 80 million person-rads of total exposure, roughly 120,000 extra cases of thyroid cancer can be expected to develop, resulting in some 6,000 deaths.

The threat posed by Chernobyl was harder to dodge. There was of course much more public outcry, and so the United Nations (UN) organized several panels of experts (literally, that is what they call them) to decide the matter. Did 1 million people die after the reactor exploded—or not? It certainly sounds like an important question to try to get the answer to right.

The UN Scientific Committee on the Effects of Atomic Radiation (UNSCEAR) itself devoted considerable time if not energy to investigating them. But when its report came out, it surprised many people.

According to these experts, the effects of the meltdown and explosion of the nuclear reactor at Chernobyl—events themselves described as one-in-a-million-year risks—were hardly measurable, and certainly not nearly as bad as people had feared.

"There is no evidence of a major public health impact attributable to radiation exposure 20 years after the accident and no evidence of any increase in cancer or leukaemia among exposed populations," concluded the UNSCEAR team.

The WHO also concluded that while a few thousand deaths may occur over the next 70 years as a result of Chernobyl's radioactive fallout, this number "will be indiscernible from the background of overall deaths in the large population group."

Coming from the UN and its "panel of experts," these views and estimates are widely reported nowadays as plain facts. But if we look past press reports to specialist publications, a very different picture soon emerges. Twenty years after the accident at Chernobyl, Dr. Eugenia Stepanova, of the Ukrainian government's Research Center for Radiation Medicine of Academy of Medical Sciences, says, to the contrary: "We're overwhelmed by thyroid cancers, leukaemias and genetic mutations that are not recorded in the WHO data and which were practically unknown 20 years ago."

For instance, in one region of Ukraine, the Rivn, 310 miles west of Chernobyl, doctors say they are coming across an unusual rate of cancers and mutations. Indeed, as well as those reports by local and presumably well-placed experts of strange and ghastly diseases, and the more downplayed but still significant studies focusing on mutated plant and animal life, there is the apparently unanswerable evidence of the countless ongoing tragedies there to be seen in the orphanages of the region.

At the children's cancer hospital in Minsk, Belarus, and at the Vilne Hospital for Radiological Protection in the east of Ukraine, the specialist doctors insisted that they were seeing highly unusual rates of cancers, mutations, and blood diseases and that they were all linked to the Chernobyl nuclear accident years earlier.

So who is one to believe? Could the World Health Organization itself be part of a nuclear conspiracy? The *Guardian* newspaper's resident energy and environment expert, George Monbiot, makes that witty suggestion in reply to claims by antinuclear campaigners such as Helen Caldicott that the WHO was biased. His blog adds: "Now, on these questions that Helen raises, I mean, if she's honestly saying that the World Health Organisation is now part of the conspiracy and the cover-up, as well, then the mind boggles. . . . If them and the U.N. Scientific Committee and the IAEA [International Atomic Energy Authority] and—I mean, who else is involved in this conspiracy? We need to know."

We can all laugh at that. But Monbiot ought to remind his readers that there is a well-known link between the WHO and the nuclear lobby. Clause No. 12.40 of the founding agreement between the WHO and the IAEA, signed May 28, 1959, at the Twelfth World Health Assembly, says plainly: "[W]henever either organization proposes to initiate a programme or activity on a subject in which the other organization has or may have a substantial interest, the first party shall consult the other with a view to *adjusting* the matter by mutual agreement" (emphasis added). The WHO may be interested in promoting the world's health (although it does not seem to be very good at it), but certainly there is a problem if it also must reconcile its reports with the IAEA's official purpose, which is "to accelerate and enlarge the contribution of atomic energy to peace, health and prosperity throughout the world."

In the case of Chernobyl, this kind of teamwork meant that the WHO's investigation was headed by scientists who had already made their names denying that the atomic bombs dropped on Japan caused the health effects that everyone else alleged they had. That was good news for both the Japanese and American governments and indeed for the Japanese people themselves—if it were true.

But clearly, there are plenty of serious people, not just conspiracy theorists, who think that radiation is deadly stuff. In fact, you need only look at another UN agency, UNICEF (the UN International Children's Emergency Fund), which is not party to any agreement with the IAEA, to find significant differences. UNICEF assessed the impact of the Chernobyl disaster in Belarus on children's health over eight years (1986–1994) and calculated the increases in deaths from various causes as follows:

- Congenital heart and circulatory diseases—up 25 percent
- Disorders of the digestive organs—up 28 percent
- Malignant tumors (cancers)—up 38 percent
- Disorders of the genitourinary system—up 39 percent
- Disorders of the nervous system and sensory organs—up 43 percent
- Blood circulatory illnesses (including leukemia)—up 43 percent
- Disorders of the bone, muscle, and connective tissue system—up 62 percent

Another investigation into the incidence of brain tumors in Ukraine by a team of academic researchers led by Dr. Yuri Orlov, a neurosurgeon in Kiev, found that this disease too had increased—by a factor of six among the very youngest infants. And similarly, Professor Alexei Okeanov and colleagues found that the real story of the radiation effects on thyroid cancer was far worse than previously expected. "It should be noted that earlier made prognosis for thyroid cancer failed, and the real picture has surpassed all expectations," they reported in a Belarus medical journal.

The same was true regarding epidemiological surveys of the effects of radiation on the people charged with cleaning up after the radiation cloud from Chernobyl had passed—the so-called liquidators. The specialist researcher, Emilia Diomina found that "small doses of radiation are statistically significant risk factors of malignant development."

And another study by Dr. Chris Busby, admittedly a Green campaigner but one with a PhD in chemical physics and a visiting professor at the University of Ulster, called "Infant Leukaemia in Europe after Chernobyl and Its Significance for Radiation Protection," in 2006 found that the effect of even low doses of radiation on unborn children (fetuses) was 100 times more serious than previous assumptions anticipated.

Finally, according to Alexey Yablokov, a prominent Russian environmentalist, Pew Marine Fellow, and former advisor to several Soviet presidents (using the standard measure for the deposition of radionuclides on soil surfaces), in a study titled "Chernobyl Consequences of the Catastrophe for People and the Environment" published by the New York Academy of Sciences:

> Chernobyl's radioactive contamination at levels in excess of 1 Ci/km2 (as of 1986–1987) is responsible for 3.8–4.4 percent of the overall mortality in areas of Russia, Ukraine, and Belarus. In several other European countries with contamination levels around 0.5 Ci/km2 (as of 1986–1987), the mortality is about 0.3–0.7%. Reasonable extrapolation for additional mortality in the heavily contaminated territories of Russia, Ukraine, and Belarus brings the estimated death toll to about 900,000, and that is only for the first 15 years after the Chernobyl catastrophe.

Contrast that with the estimates of Monbiot, from his well-insulated perch at the *Guardian.* Monbiot estimates "that so far the death toll from Chernobyl amongst both workers and local people is 43."

Forty-three? That is even less than the estimate of the WHO, which offers 58. However, Monbiot is backed in the paper not only by an editorial resolutely defending the industry but by an independent expert, one Dr. Melanie Windridge. She explains that "[d]espite Fukushima," nuclear power remains one of the safest and cleanest ways to generate power. Indeed, says Windridge, who is typical of nuclear experts but not otherwise remarkable, there had been problems at the Fukushima plant with cooling, gas explosions (not nuclear), and radiation leaks—all serious issues, but so far no one has died. No one! The earthquake and tsunami that immediately preceded the reactor problems, by comparison, killed more than 10,000 people.

In fact, she explained, the disaster actually showed how *safe* nuclear reactors are. "Reactors designed half a century ago survived an earthquake many times stronger than they were designed to withstand, immediately going into shut-down (bringing driven nuclear reactions to a halt)."

But here Windridge's mantle of academic independence begins to wear thin. According to her byline, she is a freelance science communicator. But further investigation reveals that she works as a researcher at the Culham Science Centre in Oxfordshire in the United Kingdom for an organization called the Centre for Fusion Energy, which is owned and operated by . . . the United Kingdom Atomic Energy Authority.

Fission and fusion have soaked up the bulk of UK and European Union energy subsidy and research funds for decades. Fission has given us some power, 100 tons of plutonium, and thousands of years worth of nuclear waste. Fusion has given us nothing.

Windridge concludes her independent assessment briskly:

> I do not wish to trivialise the problems at Fukushima. I dislike the radioactive waste and safety issues of nuclear fission as much as anyone, which is why I work in research into a new form of nuclear energy—fusion. It's the energy source that powers the sun and has none of the downsides of fission. Fusion will produce abundant energy cleanly and safely, but it is not yet ready. With continued political and financial support we hope to have fusion power stations by the 2050s.

Send your money in now, folks!

In her expert opinion, the energy source at fault in Japan was not so much nuclear as plain old diesel. All the problems began when the tsunami took out all the backup generators that were meant to provide power to circulate the coolant in the reactors. Loss of site power is bad news for a nuclear power plant, because as mentioned, even when a plant is ostensibly shut down, the radioactive rods at the heart of the reactor keep decaying and producing heat that has to be offset by cooling. If the rods are not cooled, the core of the reactor disintegrates in chemical (not nuclear) explosions, and great clouds of radiation are released.

And so we are back once again to the gorilla-in-the-cupboard question: How dangerous is all that nuclear radiation?

Long-lived radioactive materials have, in addition to carcinogenic effects, intergenerational effects that include the mutation of the genetic structure of life. This mutation is permanent and irreversible. Fukushima reactor No. 3 was fueled with—indeed is still full of—the mixed oxide (MOX) fuel uranium and plutonium. Plutonium has a half-life of 24,000 years, which means that it is carcinogenic and mutagenic for up to 250,000 years, or 12,000 human generations.

According to an article in the relatively specialist *New Scientist* magazine (London), 24 days after the Japanese nuclear reactors started malfunctioning, Fukushima had already produced more of the isotopes largely responsible for the almost 1 million deaths due to Chernobyl (not to mention countless illnesses and transgenerational mutations than the earlier disaster). (To be precise, about 1.7 times the amount of iodine-131 and 1.4 times the amount of cesium-137.)

And so, the magazine explained:

> Each month that passes will add two more Chernobyls of Iodine-131 and Caesium-137 to our environment, by air alone, if releases remain at current levels. The hazards, of course, will catastrophically increase if the plant becomes so radioactive that workers can no longer visit the site to keep what's left of the reactors cool.

*New Scientist* magazine added ominously that these were just the figures for atmospheric pollution—multiply them when you add in the radioactive water being pumped into the Pacific every day.

However, for nuclear experts, this spreading out of radiation across continents is a *good* thing. Both Windridge, the fusion researcher, and Hidehiko Nishiyama, of the Nuclear and Industrial Safety Agency in Japan, for example, believe that with radiation, anything that flows into the ocean "either by accident or to relieve storage problems on land" will be greatly diluted. But what do they mean? That only if you drink the entire ocean will you get a dangerous dose? Or that if you eat a fish that contains a grain of tiny but highly radioactive material, it will have been rendered harmless by virtue of having spent some time in the sea? Because, of course, radioactive particles are not actually *diluted* in the oceans; they are merely spread around. However, all this does make it hard to tell what caused whom to fall ill and die.

"And these releases are expected to continue for months, if not years. When are the 'authorities' and mass-media going to stop lying to us and belittling the extreme seriousness of what is happening at the Fukushima plant?" asked the *New Scientist.*

"Lying" is a strong word, and probably not appropriate, because the "authorities" in this case really do not seem to know or understand very much about radiation. Indeed, a surprisingly large number of nuclear experts seem not to know much either, such as the basic fact that how dangerous radiation is critically depends on whether it is experienced all at once or spread out over a long time. After all, as they say, life on Earth has evolved to tolerate a certain amount of radiation. The body has defense mechanisms that react to deformations of the genetic code in cells. But when assessing the damage caused by nuclear accidents, scientists compare one dose spread out over the 500,000-odd minutes of the year to the same dose absorbed in the one minute it takes to drink a glass of water. Experts like Windridge may not know the difference, but the human body certainly does.

Similarly, if you inhale an invisibly small particle of plutonium, the surrounding cells receive a very, very high dose of radiation. Most of the cells near the plutonium die, because plutonium is an alpha emitter.* Cells in the human body die all the time, and are replaced. The danger comes, however, when after a radioactive invasion some cells on the periphery remain viable. They mutate, and their DNA, the regulatory genes in these cells, are damaged. Inevitably, although perhaps years later, that person

* Alpha particles are a type of ionizing radiation ejected by the nuclei of some unstable atoms. They are large subatomic fragments consisting of two protons and two neutrons. Alpha particles are, however, rather feeble, unable to penetrate the protective layer of the human skin, or indeed travel very far through the air. Nonetheless, they can be very dangerous in the wrong place.

develops cancer, although the body may still fight off the cancerous cells. The same is true for particles of cesium-137, which heads for the brain and muscles; or strontium-90, which goes to bone, causing bone cancer and leukemia; or radioactive iodine, which goes to the thyroid.

In the weeks following Fukushima, the levels of cesium-137 in the village of Iitate, some 25 miles northwest of the plant, were measured at more than twice the levels that prompted the Soviet Union to evacuate people near Chernobyl. Radioactive iodine was found in the tap water in all of Tokyo's 23 city wards, some 150 miles from Fukushima. Monitors detected tiny radioactive particles from the reactor site that had spread across the Pacific to North America, the Atlantic, and even Europe. It starts to sound like the Chernobyl cloud all over again.

Windridge at least allows that leaks are undoubtedly serious. But then it is back to reassurances—remember, we are subjected to background radiation every day as a result of natural processes. There is radiation from granite rocks, there is radiation from cosmic rays affecting anyone regularly flying high-altitude routes. "And then people routinely and willingly expose themselves to large amounts of radiation for medical checks, with dental x-rays providing perhaps the highest doses, often for purely cosmetic reasons," she adds.

By comparison, nuclear facilities are famously stringent, and releases of radiation are taken "extremely seriously." Ironically, she says, it is exactly these precautionary limits that can cause unnecessary alarm among the masses. For example, take that scare about contamination of water supplies after the "problems" at the Fukushima reactor. There were recommendations for restrictions on drinking water, yet "the radiation dose received by someone drinking Tokyo water would have been less than that from moving to Cornwall and living there for a year!" adds Windridge.

Who is one to believe? Cornwall is in the southwest of the United Kingdom and full of granite rock that releases radon, a radioactive gas, into the air. But again, the relevant factor about radiation is whether it is external to the body, as in background radiation produced by rocks, or internal—as when you drink water or eat contaminated food. It makes a huge and dangerous difference.

Of course, estimates of radiation deaths are all about statistical probabilities—in other words, math, which the public is historically not very good at. We leave that to governments and their experts. Yet if radiation science is at bottom a vague science of statistical generalizations, some generalizations are more convincing than others.

In its own words, "The WHO Expert Group placed particular emphasis on scientific quality, using information mainly in peer-reviewed journals, so that valid conclusions could be drawn." Yet when it investigated the health consequences of Chernobyl, it came to a very different conclusion from the health specialists on the ground—and to one that was much closer to the political views of governments of key UN countries far away.

"The Expert Group concluded that there may be up to 4000 additional cancer deaths among the three highest exposed groups over their lifetime . . . 3–4% above the normal incidence of cancers from all causes." In sum, the WHO/IAEA estimate that the Chernobyl nuclear accident has killed 58 people and will lead, in the years to come, to "at most" an additional 4,000 fatal cases of cancer.

Specifically, the WHO uses a statistical method that dismisses any health effect from radiation below a certain level. Among those it does feel received significant doses, it predicts precisely 6,158 additional cancers in 50 years, a number that, among the 2.5 or so million cancer cases expected normally in that population over half a century, would be virtually invisible. This is because there is always a causality problem when looking at certain deaths that can have multiple causes. (Was cancer caused by radiation, smoking, or bananas?)

> Given the low radiation doses received by most people exposed to the Chernobyl accident, no effects on fertility, numbers of stillbirths, adverse pregnancy outcomes or delivery complications have been demonstrated nor are there expected to be any. A modest but steady increase in reported congenital malfor-

> mations in both contaminated and uncontaminated areas of Belarus appears related to improved reporting and not to radiation exposure.

If cancer deaths have increased by more than that—which they have—the WHO/IAEA working in tandem say it is due to "overuse of alcohol and tobacco, and reduced health care," eating unhealthy foods (like chocolate cookies), and . . . wait for it . . . worrying unnecessarily about radiation! "Radiophobia," *not* radiation, is the problem. For these experts, only one cancer seems to be directly and unambiguously attributable to the Chernobyl cloud: thyroid cancer in children. This is because there is no other possible cause for the disease. But even here, the figures are not as bad as they seem. Almost all of the cases are not "new cases" but rather are ascribable to increased reporting of the illness. Apparently, in previous years, many children had cancer of the thyroid and no one noticed. In reality, the WHO concludes loftily, the Chernobyl nuclear accident caused only nine children to die from the disease, and those deaths were really the fault of the then-communist authorities, who did not ban milk sales straight away.

Thus, the official line, if not necessarily the fact of the matter, is that only people who received high doses, such as those involved in the cleanup operations immediately afterward, should *really* be said to have radiation-induced illnesses, still a lot of people. But no; UNSCEAR reported in 2000 that just 134 liquidators received radiation doses high enough to be diagnosed with acute radiation sickness. And of those, just 28 persons died in 1986 due to radiation poisoning. "Other liquidators have since died, but their deaths could not necessarily be attributed to radiation exposure."

And finally on to the stories—and worse, pictures—of orphanages in Ukraine and Belarus, full of children with tragic and often grotesque birth deformities. Here, the WHO is particularly emphatic. Chernobyl caused no "congenital malformation" or birth defects in babies at all. "Given the low doses" of radiation the parents would have encountered, it could not have. QED. But how to explain the skyrocketing numbers of such chil-

### GARBAGE IN, GARBAGE OUT

The World Health Organization makes a key analytical error in assessing the effects of the Chernobyl cloud: thinking that in order to compare rates of mortality between people who have received different doses of radiation, it has to make a whole host of assumptions about the geographical spread of radiation and about the activities and movements of people, not to mention of livestock and crops. But since radiation clouds affect large areas erratically, the WHO figures may well be comparing rates of illness among people who although officially affected by the cloud actually suffered little contamination, with people who officially were in safe areas but in practice came into contact with radioactivity.

dren? Ah yes, but the figures have shot up all over Ukraine and Belarus. And some regions received more radiation than others. Therefore, the cause for the increase must have been something else. Maybe increased reporting, radiophobia, bananas, granite rocks, dental X-rays, chocolate cookies.

So, if that is the expert assessment, why are there massively different figures—"1 million dead," and orphanages full of children with horrible birth defects—still floating around? Because eventually the public is going to notice people actually dying. As the police say, if you have got a body, you have got a crime to investigate.

With Chernobyl, it seems we have plenty of bodies now. "At least 500,000 people—perhaps more—have already died out of the 2 million people who were officially classed as victims of Chernobyl in Ukraine." Or so says Nikolai Omelyanets, deputy head of the National Commission for Radiation Protection in Ukraine, adding that infant mortality alone increased 20 to 30 percent because of chronic exposure to radiation after the accident.

Such shocking figures would appear to spell big trouble for the WHO's optimistic report. Or they might have done, if the WHO had included such figures in its assessments, but it didn't. Why not? Because the UN has its rules, and reports have to be published in certain journals and written in English, or they will not be accepted for consideration. If a nuclear plant explodes in the United States, the WHO may get around to assessing the consequences more thoroughly.

Perhaps that is why "local" researchers, such as Omelyanets, say their information has been ignored by the IAEA and WHO. "We sent it to them in March last year and again in June. They've not said why they haven't accepted it."

An IAEA spokesman said he was confident the figures in the WHO report were correct. "We have a wide scientific consensus of 100 leading scientists. When we see or hear of very high mortalities we can only lean back and question the legitimacy of the figures. Do they have qualified people? Are they responsible? If they have data that they think are excluded then they should send it."

The WHO is not always so phlegmatic, however. Facing the problem of colds and flu, where the multibillion-dollar industries of vaccine production are involved, its response to unclear causation and small probabilities has been rather different. Take the severe acute respiratory syndrome (SARS) epidemic of 2003, for example. Then the WHO issued an emergency warning (on the weekend, no less) declaring the sickness "a worldwide health threat" after a mere *four* deaths related to the pneumonia and "another five in an outbreak of a similar infection in a province of China," albeit the two events had "not yet been definitively" linked.

Where exploding nuclear reactors produce advice to drink less milk and keep positive, cold germs produce much more dire warnings! A newly appointed UN coordinator for the avian flu outbreak of 2005 earnestly hoped that emergency measures could keep the death toll to between 5 million and 150 million. Large transnational pharmaceutical companies were asked to start preparing prepandemic vaccines. In the United States alone, $7 billion was speedily allocated to help pay for the vaccines. Yet a year later, the worldwide toll from the disease was put at . . . just 262, far less than the usual toll from the seasonal flu. On one occasion, a million people is "statistically insignificant"; on another, 262 people is a worldwide alert. Could it be because the SARS deaths happened immediately, whereas the radiation deaths were spread out over many years? Or could it just be politics?

It is a politician's job to assess future risks. That is why the nuclear industry and its regulators calculate risk on the basis of the likelihood of an accident for any one operating year. In the case of the design of the first four reactors at Fukushima, the Japanese Nuclear Energy Safety Organization estimated the "frequency of occurrence of a core damage accident is 1/100,000 or less per one year for one reactor and the frequency of occurrence of an accident leading to containment damage is 1/1,000,000 or less per one year for one reactor."

That was in 2002. Somehow, however, in 2011, four reactors were in that position, odds that must have been 1 million million million million against! Vanishingly small indeed. Funny how it happened anyway.

Or consider the risk of a major earthquake causing a nuclear plant to melt down. In Japan, the risk of a magnitude 9.0 earthquake is about 1 in every 100 years. That is clearly something that could happen at any moment. But in Japan, nuclear plants are required only to withstand tremors 1,000 times less powerful. At Fukushima, in particular, a seawall was built to cope with 23-foot waves. But the chance of greater waves has always been highly likely.

And given that only a few decades, rather than millennia, separated the accidents at Fukushima, Chernobyl, and Three Mile Island (despite prior reassuring media claims that such accidents were events that happened only once every 30,000 years), it is clear that nuclear operators and/or regulators are significantly underestimating the inherent risks associated with nuclear technology, just as they systematically understate the risk from radiation, and for much the same shameless reason.

But consider what Tony Benn, who as a UK government minister in the 1960s was in charge of the nuclear dossier and backed nuclear power, now says: "At no stage, as a minister, could I rely on being told the truth either by the Industry itself, or by my own civil servants who may or may not have known it themselves."

Disinformation on the dangers of nuclear radiation is as old as the discovery of the phenomenon itself. Even after the first nuclear bomb had been dropped on Hiroshima, General Leslie Groves, a trained engineer in charge of the Manhattan Project, was still prepared to assure the US Congress that a team of scientists had found no traces whatsoever of radiation there and that, in any case, radiation poisoning was "a very pleasant way to die." (To underline their point, the Allied powers had any journalists who tried to talk about radiation deaths deported.) Unfortunately, a month later, an Australian journalist smuggled a report out—past the US censors—from Hiroshima. This provided a significantly different picture of the day.

> [P]atients just wasted away and died. Then people . . . not even here when the bomb exploded, fell sick and died. For no apparent reason their health began to fail. They lost their appetite, head hair began to fall out, bluish spots appeared on their bodies, and bleeding started from the nose, mouth and eyes. We started giving vitamin injections, but the flesh rotted away from the puncture caused by the needle.

It certainly does not sound very pleasant . . . and in every case, the patient died.

MYTH 8

# EVERYONE WANTS TO INVEST IN NUCLEAR ENERGY

*Enrico Fermi (1901–1954) received the Nobel Prize in 1938 for "his discovery of new radioactive elements produced by neutron irradiation, and for the discovery of nuclear reactions brought about by slow neutrons." Fermi and his family used the opportunity offered by his prize to move to the United States, where he became a professor of physics first at Columbia University and later at the University of Chicago. It was there, in the squash courts under the west stand of the university's Stagg Field, that Fermi supervised the design and assembly of "an atomic pile," a code word for an assembly that in peacetime would be known as a nuclear reactor but was actually intended as the first major step in making feasible the building of the atomic bomb. Today, a plaque at the site reads: "On December 2, 1942, man achieved here the first self-sustaining chain reaction and thereby initiated the controlled release of nuclear energy."*

*Enrico Fermi explaining the first self-sustaining nuclear chain reaction. Today a real danger exists that collapsing investments in nuclear power could lead to another kind of self-sustaining chain reaction. (US National Archives/Argonne National Laboratory)*

"THE NUCLEAR INDUSTRY WANTS HELP TO BUILD new plants," announced US newspapers recently. But that is not quite right. The nuclear industry *needs* help to build more plants. We all should chip in. The most outspoken supporters of the nuclear industry in the United States are Republicans, although Democrats are pretty cozy too. Many prominent Republicans, such as John McCain, senator from Arizona, called in 2010 for no fewer than 45 new reactors to be built in the next 20 years. Senator Lamar Alexander of Tennessee outbid him, with a call for 100 more reactors by the same date.

But the trouble in the United States is that no one wants to invest in the technology. It just is not profitable. From 2001 to 2007, during what should have been golden years for the industry, the price of electricity on the wholesale market in New York crashed by half, from its high of just under $100 a megawatt-hour (MWh) to a paltry $47. Even with optimistic calls to tax nuclear energy's competitors at $20 per MWh, companies investing in nuclear power would be unable to make a profit. This is where the go-go financing, special investment vehicles, and zany new schemes of all sorts come in. Yet even with states and central governments in the United States and elsewhere coming up with multibillion-dollar loan guarantees offering free money to investors, the industry is failing to keep up with costs. The reason: Today new nuclear plants are hugely expensive, and the investment environment is not what it used to be. Privatization, cheap natural gas, tightening public purse strings—all have affected the nuclear calculation. These days, utilities have to lobby repeatedly for their subsidies, just like everyone else—and they still do not get all they want. That is why some have loudly dropped their plans to construct glossy, new $6 billion nuclear reactors. In April 2009, for example, AmerenUE, an electricity company based in Missouri, did just that when efforts to get the state legislature to set aside its long-standing ban on subsidies to work in progress failed. A similar state refusal to pick up the bill for extra costs prompted the Florida Power and Light Company to refuse to proceed with planned reactors at Turkey Point in 2010. (It had wanted an extra $1.27 billion.)

Unfortunately for the nuclear renaissance, as one anonymous member of the New York State Public Service Commission, a body whose job, among other things, is to fund new power stations, put it: "It's almost unfinanceable in today's environment." So today, the search for ways around the obstacles has become increasingly urgent—with the go-to being accounting tricks that make Enrico Fermi's nuclear math look simple. As a result, nuclear finance assets are increasingly being used as supports for unsafe and unsustainable financial packaging. It is a strategy that amounts collectively to a nuclear asset bubble capable of bursting with consequences every bit as disastrous as those following the subprime mortgage bubble in 2009.

Generally, large projects are financed through a combination of debt (borrowing from banks) and equity (self-financing from income). Perhaps surprisingly, to outsiders, it usually makes more sense for large companies to borrow money than to use their own.

One reason is that by financing investment from equity, the company is asking shareholders to defer sums that could have been paid immediately as dividends. To compensate shareholders both for this loss of dividends and for the risk that is being taken with their money (at the very least the risk that the project might never make its expected return on investment), the cost of equity—that is, a company using its own money—is therefore generally higher than the cost of debt. Another advantage of borrowing is that once large loans have been arranged at low interest rates—perhaps with government support—the money can then be lent out at higher rates of return. Nuclear construction companies are often a kind of shell vehicle in which finance experts play with huge sums of money and different rates of interest.

Unfortunately for nuclear financiers, sometimes banks do wonder why a company is not prepared to risk its own money and may refuse to lend. In such cases, replacing borrowing with equity is almost the only option for them. Almost. Because into this impasse steps the national government, offering loan guarantees. All financial problems are magically solved! No one stands to lose money—except the public, of course. It is particularly revealing that in recent years, six of Wall Street's then-largest investment banks—Citigroup, Credit Suisse, Goldman Sachs, Lehman Brothers, Merrill Lynch, and Morgan Stanley—informed the US Department of Energy that they were unwilling to extend loans for new nuclear power plants unless taxpayers shouldered 100 percent of the risks. Utilities successfully lobbied for 100 percent coverage of debt at up to 80 percent of the project cost.

When taxpayers in the United States were made to take all the risk of lending for new houses—the subprime crisis—the short-term aim was to increase the number of loans made for new homes and to give a boost to the construction industry. We all know the long-term results. Government agencies like Fannie Mae and Freddie Mac made hundreds of billions of dollars of evidently risky loans and then had to be bailed out by the national treasury, with US taxpayers and international creditors eventually picking up the bill. The expanding ripples of debt would eventually sink economies around the world. Actually, even as long ago as 1975 to 1979, the boom period for the nuclear industry that was brought to an end by Three Mile Island, nuclear power financing had already become a classic asset bubble. Nuclear power had for all intents and purposes lost its credibility at that time—not due to health, safety, or environmental costs and risks but due to fantastic cost inflation. Right through the value chain, starting with a spiral of building and operating costs for nuclear power plants, nuclear power had become an overheated asset not unlike certain other things today—such as Internet shares, or packages of subprime loans. . . .

But old habits die hard. The US Energy Policy Act (2005) tried to jump-start the nuclear construction industry by, for example, offering tax credits for electricity produced (simply handing out free money to the companies); paying interest bills if a nuclear plant takes longer to build

### FEED ME!

What a lot of the new political clients for nuclear power do not seem to realize is that the industry likes starting projects, getting the funds in place—and then never actually finishing them. Worldwide, only half of the contracted reactors ever get finished. Consider the situation in the United States, the world's largest economy, for example. At the end of the millennium the George W. Bush administration made a concerted effort to revive the nuclear sector under a nuclear power program unveiled in 2002. The US Department of Energy was to launch cooperative projects with industry for new generation reactors. To save potential investors the trouble of obtaining regulatory approval for these doomsday machines, the plan was to issue official approval in advance covering sites, construction methods, and operating licenses. (This was the so-called early site permit process and combined construction and operating license.) A large number of reactor designs exist, but many are not far advanced, do not have regulatory approval, and have limited prospects for being ordered.

To help interested companies obtain the paperwork, a half billion dollars or so of grants was made available. Various businesses soon combined into two organizations: NuStart, launched in 2004, and another group led by Dominion. Yet although both NuStart and Dominion insisted that they intended to pursue the licensing process, neither group committed to building a new plant, and no reactor orders have been placed.

The chief executive of Dominion, Thomas Capps, put the initiatives of the companies in perspective in May 2005 when he stated succinctly: "We aren't going to build a nuclear plant anytime soon. Standard & Poor's and Moody's [the debt-rating agencies] would have a heart attack. And my chief financial officer would, too."

than originally estimated (which happens almost every time); and offering federal loan guarantees for 80 percent of the total project cost. This last enabled companies to borrow money at far lower rates. It is as if a family was able to go to a bank and ask the skeptical manger for a mortgage, saying that if by chance it defaulted, the government would pay back the bank. Of course, such arrangements are very convenient for entrepreneurs with spotty credit histories—but they can be ruinously expensive for governments.

In expecting public funds to take financial risks, the nuclear power industry is aided by the fact that a simple definition of "subsidy" is difficult to find. The usual assumption is that a subsidy is some kind of direct cash payment offered by a government to an energy producer or consumer. But this is just one way in which governments can stimulate the production or use of a particular fuel or form of energy. There are plenty of other, much cleverer ways. For example, in order to help energy producers, governments can rig markets to either raise the prices customers must pay or engage in a vast array of measures that reduce their costs. The so-called feed-in tariffs that encourage entrepreneurs to invest in solar panel arrays or wind turbine parks, and which work by obliging electricity grids and hence customers to pay many times the going rate for the energy produced, are an example of the former kind of subsidy. This trick is very popular with governments, which typically take a cut of the artificially inflated energy costs as well as numerous other side taxes. By 2010, feed-in tariffs were being used by governments across Europe and in Canada, China, and Israel and (at the state rather than at the federal level) in the United States (notably California) and Australia.

Unfortunately, the problem with this cunningly concealed energy tax is that feed-in tariffs soon become fantastically costly to consumers. In Germany, the cost of subsidies for solar power alone was expected to reach no less than €46 billion by 2030 (€250 euros per MWh), while in heavily indebted Spain, the uptake was so high that the government was forced to backtrack on its commitments (particularly on the tasty feed-in tariffs) after investments had been made.

Britain, too, is heavily indebted, but that has not stopped the climate change bandwagon there. Despite increasing public skepticism, an energy bill was passed that includes some really rather extraordinary figures, figures that other citizens around the world whose governments are busy drawing

## MORE US AID

In the United States, tax breaks and subsidies included in legislation like the American Power Act include such things as government insurance against regulatory delays and federal loan guarantees that both enable constructors to make money out of the difference between the normal price of borrowing and their privileged access to government fund), which often runs to tens of billions of dollars. More generously still, not only does the US government offer cheap loans to nuclear companies; it also picks up the losses when projects go bad.

### Production Tax Credits

In order to make electricity generated from new nuclear power plants competitive with other sources of energy, an $18/MWh tax credit would be paid for the first eight years of operation. According to the US Energy Information Administration, this subsidy alone would cost US taxpayers $5.7 billion by 2025.

### Loan Guarantees

The Congressional Research Service estimated that the taxpayer liability for loan guarantees covering up to 50 percent of the cost of building six to eight new reactors would be $14 to $16 billion.

The energy bill passed in 2007 gave the US Department of Energy a budget of up to $18.5 billion for 2008–2009 for loan guarantees covering nuclear plants. But this sum was enough for only one or two projects, so in February 2010, in its 2011 budget, the Obama administration approved an increase in the amount available for loan guarantees, from $18.5 billion to $54.5 billion. That was perhaps enough for 12 new nuclear piles.

But that's not all. The rather grand-sounding American Clean Energy Leadership Act of 2009, enthusiastically promoted by the US Senate Energy and Natural Resources Committee, authorizes unlimited loan guarantees, while the ultra-green but endangered Kerry-Graham-Lieberman framework for climate legislation adds some more nuclear loan guarantees. Another recent bill calling for 100 new nuclear reactors features new subsidies that could leverage between $100 billion and $1 trillion in loan guarantees. Finally, the industry itself is on record calling for a 1 percent subsidy, which on its own would result in extra costs to the public in the amount of $1 trillion.

up ostensibly green new energy acts should note. They appear in documents produced by the Department for Energy and Climate Change in the United Kingdom (that is the latest name, upgraded from plain old "Department of Energy") that were published in July 2009. The documents shamelessly say that the total cost of the increase in energy prices resulting from carbon emissions reduction and renewable energy polices is going to be around £5.7 billion per year in 2010 and is set to rise, under existing and proposed policies, to over £16 billion per year in 2020. By that time, the energy department gloated, it would constitute around 2 percent of entire UK taxation—a hike of 25 percent in the basic rate of tax for most people!

Yet it seems as long as the tax is for renewables, no one in the United Kingdom minds too much. But that is the power of greenwash. If beer prices there go up, there would be furious complaints, but if water meters (rare in the UK) are made compulsory, the gas tax is ramped up again, and homeowners are told they must install double-glazed, plastic windows, the famous British stiff upper lip comes into play. Governments know that people will accept financial pain in the name of saving the planet. Tax hikes for nuclear reactors would prompt a different reaction; hence the hidden cash trail.

In truth, the subsidy slush funds created for the energy industries are enormous. In Britain, a country with only a modest annual gross domestic product, the government estimated the costs of implementing its green energy plan at £324 to £404 billion. Loads of money to build nuclear power stations, plus a bit for the grand tidal energy schemes of the German corporation E.ON. Mind you, according to Carbon Market Data, E.ON was in 2008 the second-biggest emitter of carbon dioxide in Europe (over 100 million tons). But the Greens still want to help them.

The nuclear power industry welcomes free money but, as we have seen, its particular form of power generation is still not economical. So, even now, nuclear operators can often be found in regulatory circles demanding higher and higher carbon prices (i.e., the levy on nuclear's competitors), sometimes up to $75 or even $150 (€50 or even €100) a ton! Électricité de France (EdF), with plans to build four new nuclear reactors in the United Kingdom, specifically warned the British government that it should set this sort of price for carbon permits if it wanted the corporation to "invest" there. (Given EdF's pitiful financial situation, its threats may not have carried much weight.)

Back in December 2009, at the time of the Copenhagen climate conference (when many people expected an international agreement to make penalties for producing carbon dioxide mandatory on all the counties of the world), European exchanges were trading carbon permits* at the price of £14 a ton. Today, the permits are almost worthless and the carbon exchanges are closing. Fortunately for the nuclear industry, it has already managed to make a tidy windfall profit selling the free permits it received early on, in honor of its supposed "near zero emissions" status.

And despite the debacle that the Copenhagen conference turned out to be, and the fact that now any international agreement on mandatory carbon prices seems hugely unlikely, politicians in Britain, Australia, and California have all set their own "carbon" targets. Because of course, governments will not really turn their backs on their nuclear friends. They simply move on to offering other subsidies instead. One such new sweetener for nuclear is to be counted as a "renewable" energy source. Thus rebranded, it could take a share of new green taxes, such as the United Kingdom's renewables levy, which the government imposes on electricity bills by obliging electricity providers to purchase renewables. For the UK nuclear industry, such a change in status would be worth up to £300 million a year, or £3 billion over a decade.

The bottom line is that for all the spurious technicalities, energy politics is very simple: It is about money. Whether nuclear power is carbon free, mostly carbon free, or partly carbon free has no lingering significance to creative financial industry players. The atom is always friendly because massive reactor investment proposals will always generate good business, directly or indirectly.

---

* These are official permits that entitle them to use technologies resulting in the dreaded carbon dioxide.

The nuclear industry's interest in climate change is all about money. International energy corporations (such as E.ON and EdF) cheerfully build *both* nuclear power stations and green energy parks, thereby enjoying two slices of the "low-carbon" pie. Yet for the industry, since its reactors are up to nine times more expensive to build than equivalent gas-fired plants, often the only way to tempt private investors to build them today is to persuade governments to step in and tax competitors, such as coal and gas, for their emissions of carbon dioxide.

Of course, another way to ramp up costs for the coal and gas industry is to change the rules they operate under. Despite its own love of subsidies, the nuclear industry is first to complain about those enjoyed by its great rival: coal. In 2010 for example, moaned the World Nuclear Association, some €3.2 billion in coal subsidies was to be "handed out" by six EU countries: Germany, Hungary, Poland, Romania, Slovakia, and Spain, all in order to safeguard 100,000 or so evidently worthless jobs! But the Eurocrats of Brussels have taken the industry's point, with mining subsides steadily being phased out by a (selectively) competition-promoting European Commission. The Commission does not seem so bothered if the nuclear industry quietly accepts all *its* new subsidies, both European and elsewhere, such as the $199 billion gift that came about as a result of changes in tax rules related to decommissioning under the Energy Policy Act in the United States.

Another tool the nuclear industry uses to distort the energy playing field in its favor is the issue of new, supposedly deadly forms of air pollution, such as sulfur and nitrogen oxides. As with carbon taxes, the European Union has been at the forefront of research supposedly demonstrating the high external costs of fossil fuels.

For example, back in 2001, ExternE, a very expensive if not very substantial US and European study of the external costs of various fuel cycles, focused on coal and nuclear and claimed to have put "plausible financial figures against damage resulting from different forms of electricity production."

The external costs considered were things like the effects of air pollution on human health, crop yields, and buildings as well as occupational disease and accidents. The 2001 report excluded effects on ecosystems and the impact of global warming, but these were added later despite the acknowledged and indisputable range of uncertainty in quantifying and evaluating them. In other words, garbage was put in and certainly garbage came out.

The most environmentally minded thing about the ExternE report was that it basically recycled an earlier 1999 European study. This had put environmental damage costs from fossil fuel electricity generation in the European Union for 1990 at $70 billion while allowing the nuclear industry off scot-free by making calculations of the likely effects of radiation every bit as optimistic as those of the United Nations' World Health Organization, which, as mentioned, is contractually obliged in such matters to follow the views of colleagues at the International Atomic Energy Authority. Once this marvelously low risk from nuclear energy emerged, it was contrasted with aggressively high estimates of radiological impacts from mine tailings (estimates later easily shown to be exaggerated). All this fitted in very well with the political desires of various EU governments to cut back on subsides to their mining communities. As for the nuclear industry's external costs related to the never-started and never-ending tasks of waste management and decommissioning, well, these were not included, since officially they have already been costed and paid for by the industry.

The end result, not so much of the research but of the political maneuvering, was a report that appeared to demonstrate that, in cash terms, nuclear energy has a bare tenth of the external costs of nasty old coal, which made it cleaner even than hydro. Only wind showed up better than nuclear.

So what does all that prove? Only that, as with the entire international climate science process (with its five-yearly International Panel on Climate Change reports, etc.), science is as easily co-opted by politicians as politicians themselves are by lobbyists. But no matter! If these external costs could, in fact, be included, the price of electricity from coal in Europe would increase 50 percent, that from gas would increase around 30 percent, and (at last!) nuclear energy would be (almost) competitive.

The United States was an original coauthor for the European study but, as with the Kyoto Protocol, it soon dropped out. But not because the US government objected to the ill-judged political efforts to skew the energy markets. Rather, it wanted to do the same thing its own way. The result was that, in October 2009, a US National Research Council report claimed a total of $120 billion in "hidden" external costs of energy production in the United States for the sample year studied of 2005. Like the European effort, the figures mainly relied on estimates of health damage and excluded the effects of climate change. Among energy generators, practically all the costs were attached to the coal industry, while all the prizes went to the nuclear industry.

Such then is the politics of nuclear finance. But politicians eventually come into contact with economic reality, and once that happens even lobbying is less effective. The bottom line is that the economic and financial situation for nuclear energy has always been starkly different from other energy sources. The costs just keep rising. The nuclear industry crawls up a wall of worry and falling subsidies. This is the near-term outlook in all of the old-nuclear countries and a process that must also affect atomic energy in the emerging economies. In old-nuclear Europe, as many as 40 to 50 reactors will have to be taken out of service in the next two decades—if that is possible—at an average cost that even nuclear optimists put at $2 billion or more per reactor.

This cost wall is a daunting prospect—almost a kiss of death for the illusion of atomic energy delivering cheap electricity—and has spawned a number of technological, financial, and legislative reactions from the industry and the governments that support it. One key "strategy" has been quick, cheap reactor lifetime extensions but, again, the Fukushima disaster, where one of the stricken reactors was within days of its fortieth birthday, has firmly focused world attention on the danger of old reactors and their ancillary equipment.

Taking reactors out of service, fully dismantling them, operating full "Safestore" confinement, and guarding their complete radiological inventories is probably one of the greatest non-wartime challenges these countries will ever face. To date, almost all civil reactor retirements have been on a piecemeal basis, often forced due to accidents, with no organized and structured industry-wide norms, guidelines, and industrial preparedness.

The era of cheap, discreet, and effortless reactor lifetime extensions, at least in countries where public opinion is vigilant, is coming to an end. If it does, the costs facing nuclear power become ever more daunting, and electricity prices will have to rise even faster and further.

At the same time, building new reactors anywhere is becoming much more expensive. For 2010–2011, the US Energy Information Agency reports that reactor building costs are growing at *37 percent a year* from a base of $5,339 per kW installed—flamboyantly ignoring the cost of advanced technology gas-fired power plants. Since 2005 worldwide, reactor construction costs have at least doubled, but the doubling time is shrinking, and it is entirely possible that they may double again by 2014, meaning that capital costs could attain the stupendous height of more than *$10,000* per kW installed. Of course, it can be (and in nuclear circles it sometimes is) argued that this is not so very much more than the highest-tech offshore wind farms or some of the more experimental renewable power systems, such as biogas-powered fuel cells, but the implications for mainstream power remaining associated with nuclear energy are economically prohibitive.

Conversely, fossil-based electricity-generating costs have declined in price, in real terms, a difference that shows up in stark relief in the capital costs. For instance, typical 2011 costs for a gas turbine–based power plant are around $700 per kW compared to as much as $7,000 per kW for the third-generation European reactors being built in Finland and France. For the nuclear industry, the sad fact is, despite rising fossil fuel prices, especially of oil and coal, fossil-fueled power plant construction costs have generally fallen for the last 20 years in real terms while nuclear power plant capital costs have only risen.

Curiously, or perhaps perversely, the uneconomical reality of nuclear power is a major reason for the ostensible privatization of the nuclear industry. Cynics will say this is just another example of privatizing profits and nationalizing losses, but it still demands a bit of explanation. As with many other formerly nationalized industries in a wide range of countries, industries and services ranging from the post office to sewage treatment, privatization has always targeted the juicy bits and shied away from social responsibilities. As a result, the lucrative parts of national economies are sold off to the private sector while an often cash-strapped state is left with the unprofitable remains. Today, in the old-nuclear countries, many areas of nuclear power are being privatized: the upstream fuel fabrication industry, fuel reprocessing and waste handling, and reactor decommissioning.

The desire of the nuclear power industry to hold on to its profitable segments explains why Germany's biggest nuclear power producers (despite at first having fiercely resisted the coalition government's nuclear levy or special tax on nuclear generation and distribution) fully welcomed Chancellor Angela Merkel's May 30, 2011, decision to completely abandon nuclear power by 2022. As long as the industry was assured it would not have to pay the real costs of decommissioning plants, it soon realized on which side its bread was buttered. Put another way, producing electricity and seeking special subsidies was yesterday's game. High-tech nuclear services was a much more promising one for tomorrow.

To be sure, the state is still everywhere in supporting the nuclear industry. In countries where the industry is semiprivatized, the state will wield its golden share* in the shadowy, government-friendly nuclear power and engineering conglomerates and consortiums. This is typified by Areva of France and its few main rival engineering and power plant builders, led by the Japanese-US Toshiba-Westinghouse and GE-Hitachi groups, KEPCO

* That is, use special privileges and rights as an investor; literally a golden share is a nominal share that is able to outvote all other shares.

of South Korea, RusAtom of Russia, and the China Guangdong Nuclear Power Group.

Meanwhile, the canny Germans have beaten a full retreat from nuclear engineering: Siemens of Germany, a longtime reactor-building and nuclear equipment specialist and one of the world's earliest fully integrated nuclear engineering companies, was sold, with all its subsidiaries, to Areva in early 2011 for the unimpressive—indeed rather derisory—sum of about $2.25 billion. By the middle of 2011, the financial health of the two Japanese-majority Japan-US conglomerates was seriously weakened by their debt loads and the dearth of firm orders.

Even before Fukushima dealt a powerful blow to fragile corporate confidence in the nuclear sector, many businessmen realized that the writing was on the wall, in large part because of the high gearing, or debt-to-equity, structures of the remaining major nuclear plants and engineering conglomerates. Any coming rise of interest rates would just be the final nail in the coffin.

The increasing necessity to prop up nuclear power financing with complex and fragile products, such as the imposing-sounding structured investment vehicles beloved of the new generation of computer-savvy financiers, helps explain why political leaders like Barack Obama and Nicolas Sarkozy have regularly called for new global financial facilities to aid nuclear power. The favored tool has been a special nuclear money pot, to be managed by the International Monetary Fund (IMF), with the mission of supplying easy financing for sales of nuclear plants and technology in the new-nuclear South. However, this idea has almost certainly perished forever, along with plenty of other illusions, in the heat of summer 2011's global debt-and-finance crisis.

In particular, proposals made in 2010 by the former head of the IMF, Dominique Strauss-Kahn, just before his abrupt departure from the world financial scene (following his arrest in New York for allegedly assaulting his hotel maid) included the use of sovereign debt–linked instruments, currency and interest rate swaps, and similar components in financial packages for so-called green nuclear energy in lower-income countries of the new-nuclear South. This gives an idea of their scattershot approach, using every possible financial engineering trick to transform the corporate borrowing required to develop and own nuclear plants into national and sovereign debt. Nuclear sector examples of this metamorphosis already exist; consider, for example, the sophisticated financing used to laboriously complete the sarcophagus over the ruined Chernobyl reactor. The global value of paper assets riding on this project is impossible to define, because these assets change in value with movements in the stock market and bond yields (on Ukrainian debt), but it is likely at least 100 times the estimated cost of the entombment. The cost of the "Sarcophage" itself, as noted in Myth 3, is also unknown and changing, but estimates place the final cost, by 2012 or 2013 (when it is supposed to be finally completed), anywhere between $1.2 billion and $2.5 billion.

The Chernobyl cloud has indeed had a silver lining for the nuclear finance world. Similarly, other hoped-for projects in the South also play with debt, creating incredible multiples of risk. Typical schemes extend to four reactors per project site, adding up to a project price tag as high as $25 billion. This amount is supposed to be financed by borrowing. (As discussed, debt is always preferred to equity, or using your own money.) Thus, the amounts of paper floating on the underlying asset—one single large nuclear complex—can very easily surpass $2.5 trillion. This sum is comparable to the total national public debt of major world economies, such as Italy or France. The almost open-ended and certainly extreme financial risks of these house-of-cards financing deals are very plain to see.

Just like the US subprime rout, the related financial leverage, which is ironically called risk spreading, includes participation by major insurance companies. If the hoped-for reactor order falls through, if the sovereign borrower defaults, if the local economy descends into riot and anarchy, or in any number of other cases of systemic shock, the insurers, as well as the lead banks in the syndicated loan arrangement, will all face disaster. Given the fallout from the US subprime crisis, this new potential rival for future financial disaster is the worst possible thing at the worst time.

Of course, most of the finance industry and its armies of traders and brokers will probably survive a nuclear financial meltdown, as most survived the US subprime crisis, but the clear losers will be humble electricity consumers and basic-rate taxpayers. In the new-nuclear South, the collateral damage could be much higher, causing a repeat of the Third World debt crisis of the 1980s, during which time countless millions of people were thrown into absolute poverty.

As we saw in Myth 5, the stampede to nuclear power led by China and India has led to its adoption by as many as 20 developing countries, from Ghana and Sudan to nearly all major Arab states and through Asia, from Mongolia to Bangladesh. This was the situation as of March 11, 2011, but since the Japanese disaster, the Arab Spring revolt, and gathering storm clouds for world monetary and finance stability, the trend toward new and ever-larger reactor building projects and proposals in developing countries is almost certainly waning.

The massive recentering of nuclear industry sales effort to the South required complex, fragile, and delicately balanced financial packages. In many cases, these sales come with 30- to 50-year credit-based financing packages with the potential to rapidly shift into millstones of debt under anything but a continuation of strong growth of countries' economies and national budget and trade surpluses.

Perhaps ironically, leaders of the three key oil exporter countries of the Arab world—Saudi Arabia, the United Arab Emirates, and Kuwait—are proponents of the oil-saving rationale of nuclear power. These nations have signed up for massive reactor projects since 2009, financed by elaborate credit-based floating financial packages with linked tradable instruments. The odd claim is that nuclear power will somehow stretch the oil age, by reducing domestic oil demand growth, and enable these countries to keep exporting oil longer, even as their production diminishes. In all cases, nuclear costs are externalized to future budget rounds, indeed to future generations.

For the new-nuclear emerging and developing countries, many of which suffered heavy economic damage for as long as 15 years after 1985, during the Third World debt crisis, the long-term sustainability of the financial packages and the uncertain performance under stress-test scenarios of falling economic growth, lower export revenues, budget and trade deficits, and weakening national currencies point to only one conclusion: *economic meltdown.*

AFTERWORD

# WHERE NOW?

## SEARCHING FOR SOLUTIONS TO THE RIDDLE OF THE INSATIABLE DEMAND FOR ENERGY

**BY NOW IT MUST BE CLEAR TO EVEN THE MOST NU-** clearphile reader that the debate over the role of the friendly atom in the modern world is not so much economic as political. Because the subject involves so many technical and financial complexities, it is easy to convince the wider public that such matters are best left to "experts." But as this book has shown, energy decisions are not being based on such technicalities. Instead, politicians and powerful interest groups are making them on the basis of short-term political calculations, while a host of media experts and public relations experts are employed to direct the public debate in directions that lead nowhere.

Even reasonable discussion about the enormous, if not insurmountable, technical, geographical, resource, and economic challenges posed by plans to supply as much energy from renewables as we get from fossil energy sources has been in practice *verboten.* In the United Kingdom, for example, in 2008, a climate change act was passed with scarcely any national debate, even though it has profound, indeed unprecedented, consequences for the country's future. As Gordon Hughes, an energy economist at the University of Edinburgh, argued in a rare contrarian thrust to the prevailing consensus, the British government target for generating electricity from renewable energy sources involved capital costs nine or ten times as great as those required to meet the same demand by relying on conventional power plants. Thus, the extra investment required for renewable energy would have to be diverted from more productive uses in the rest of the economy.

Similarly, when health and environmental benefits clash with business and commercial ones, the outcome has little to do with cutting-edge science. For example, in September 2011, when US President Obama canceled sweeping new restrictions on low-level ozone pollution proposed by the US Environmental Protection Agency and backed by a committee of independent experts, he was obliged to do so not on the basis so much of rational science as of gloom-laden warnings about the costs to business. Despite its earlier pledges to base decisions on science, the White House now felt that the research behind its initial decision needed to be revisited, and so a new standard would have to wait to be issued . . . after the next election.

The fact is, in a world where industrial and technological development is the right of all nations, not just a few, it is difficult to have a real debate about world energy needs. Instead, a crude and superficial picture of "modernity" is superimposed on supposed national prerogatives and priorities. India and China are cases in point. Both countries are devoting a huge and morally indefensible proportion of their national resources to creating a road network (highways, bypasses, bridges, parking) to service the lifestyles of a tiny car-owning elite. Cars, of course, gobble energy. But so does making them, and so does building and maintaining roads, not to mention the car's enormous energy implications in terms of pollution and changed land use. All of this development is misrepresented as a kind of democratic extension of western lifestyles to the developing world. But, in reality, it is the skewing of opportunities and resources not only away from the 1 billion or so people in the world who lack food or basic sustenance, but indeed away from almost all of the 7 billion people on today's energy-challenged planet, toward a handful of huge corporations and their national backers.

Even on the narrow issue of electricity generation, arguments and payments will continue long after everyone has given up on solar and wind—energies that have never and will never play a significant part in the world energy mix. (Figures that make the renewables option look significant invariably cheat, sometimes by adding in hydroelectric power, sometimes by talking about "installed capacity" rather than actual usable energy provided.) Political friends of nuclear energy will shed crocodile tears over the fact that new renewable sources of energy are not only cripplingly expensive but also simply unable to provide large amounts of energy, reliably, on demand. They will brighten, though, in explaining that "the gap" has to be met somehow, and that if fossil fuels are ruled out, the only technology left seems to be nuclear. Yet, of course, there is an extremely important alternative solution to energy shortages: Use less of it and make better use of that which we do have.

So why, if there is this alternative solution, is it not being offered? Of course, the answer is political. For all the talk about the liberalization of energy policy, the privatization of electricity, and so on, we still allow huge energy corporations to rearrange the world as they see fit. Today, how energy is used, and how much is required, essentially rests on large-scale strategies that only governments—and, in some cases, only transnational organizations—can implement. At the national level, for example, huge amounts of energy are used to move people around—from their detached homes to their town-center offices, to their shopping centers, to their schools—rather than to create networks of largely self-contained, fully functional communities. Huge amounts of energy are used to heat and air-condition buildings rather than build them so that they are more energy efficient. One evergreen energy-saving example: The European Union has calculated that widespread use of fiberglass roof and wall insulation in its area alone would save almost 100 million tons of oil annually, equivalent to the annual energy consumption of buildings for 90 million inhabitants.

For truly, there exist plenty of simple energy-saving strategies, from making such insulation free to making fuel-guzzling cars prohibitively expensive. But such solutions are technocratic, politically unfashionable, and, most crucially, require joined-up thinking. Governments prefer to point their voters at inefficient, even energy-wasteful, strategies, such as using light bulbs filled with highly toxic chemicals made in the Far East that require taking (by car no doubt) to special recycling points at the end of their lives. Indeed, the entire recycling movement has repeatedly been shown to be vastly more energy-greedy than is justified by the materials reclaimed. Recovery of metal may make sense, but the rest of what we think of as recyclable material should ideally be fed to one of the new large-scale waste plants. But such plants cost governments money, so they prefer to offer fake Do-It-Yourself solutions to gullible publics.

Or take global trade. World shipping uses at least 4 million barrels of oil each day. There is simply too much stuff being moved around. In energy terms, it would make sense to combat this market-driven strategy where everything is outsourced—from food production to high-tech consumer goodies—(always seeking to save on wages and environmental protection), and actively promote local sourcing and sustainability instead.

No surprise though, that at present national strategies and decisions are going in the opposite direction. Planning, for example, for changing global freight moves is hopelessly unconnected. Ships are owned by private international companies. Ports typically are publicly owned but controlled by local municipal governments, and railroads serving ports are usually private or run by national governments. Today it would seem to be easier to get a person to Mars than to organize world freight efficiently. Yet meeting such challenges is the true, long-term solution to global energy imbalances.

A specific example, from the United States. Freight moved by water uses 586 kilojoule (kJ) per ton per mile; in contrast, freight moved by heavy trucks uses six times as much energy: approximately 3,850 kJ per ton per mile. The planned expansions of ports in Canada and Mexico will result in a huge increase in freight arriving by road in the United States. It's a no-brainer in energy terms: these ports should not be expanded; rather, the US ports should be.

America's highway network carries over three-quarters of the nation's freight, as measured by tons shipped. In western Europe, the figure is two-thirds. That may not sound very much better, but the amount of energy involved is certainly enough to bridge the difference between keeping nuclear energy and abandoning it.

But above all, in the United States, energy politics is directed by the nation's love affair with the car. Americans drive cars, motorcycles, trucks, and buses about 3 *trillion* miles per year. The gasoline and diesel used makes up 84 percent of all of the energy used in the country for transportation. Clearly, cars aren't going to disappear tomorrow. Yet even *what* we drive makes a difference: A BMW 318i will get 47.9 miles per gallon, but a Cadillac CTSV manages just 18.5 mpg, while among family cars, a Volkswagen Golf 1.6 diesel will cruise along doing 62.8 mpg, where a Subaru Impreza 2.5 guzzles 23.7 mpg. Just switching to fuel-efficient vehicles of the same type could reduce America's total energy needs by 20 percent!

In a purely free market, philosopher and economist Adam Smith might be right in predicting that individual choices will collectively amount to the best possible use of price and resource information. But in the real world, there are no truly free markets, only complex webs of interlocking special interests, erratic government regulations, and competing national priorities. Individual decisions cannot make a difference here, no matter what the Green campaigners tell us. Yet on one traditional item of faith at least they do have a point: If lots of people save a little bit of energy, then, overall, considerable efficiencies can be achieved. The fact that this is not mere wishful thinking was illustrated in the aftermath of the Fukushima disaster, when the Japanese government called for a 15 percent cut in electricity use in Tokyo and the surrounding region, saying it was absolutely necessary to avoid blackouts and power shortages. The Japanese fell-to with characteristic social zeal. But then *setsuden* ("power saving" in Japanese) has long been part of the Japanese political vocabulary, and the Japanese environment minister, Ryu Matsumoto, was only reinforcing a long-held belief in the strategy when he said that the reduction would not be a temporary measure but was intended to change people's lifestyles forever.

Traditional Green exhortations to turn off lights when not in the room, switch off computers at night, nudge thermostats, and arrange car-sharing were all part of this Japanese crisis agenda, but so were larger-scale tactics for businesses. Office complexes switched off their air-conditioning entirely and told workers they could dress in shorts and shirt sleeves. Commuter trains were made to run slightly slower. Working hours, "family friendly" or not, were adjusted to use energy more efficiently—the Casio Computer Company even changed weekends for employees to

Sundays and Wednesdays to help ease the power load on the electricity grid during the workweek. In what was surely the final triumph of Green thinking, Hitachi and Kyocera began to grow curtains of plants to cover their factory walls and shade their windows.

So what's all that prove? Very simple. At present, nuclear power meets just 2 or 3 percent of world energy demands—not nearly enough to justify its huge investment and risk. Realistic and straightforward planning strategies can reduce energy consumption far more without putting lives in danger and laying waste to whole regions at a time.

# GLOSSARY OF KEY TERMS AND BACKGROUND BRIEFING

**Areva** Areva is a French multinational corporation whose primary activity is to construct nuclear reactors for the French electricity utility, EdF. Areva is unique worldwide in that its business covers all the stages of the so-called uranium cycle, from mining and enrichment to the eventual storage and disposal of nuclear waste. This broad capacity reflects the corporation's origins in the merger (in 2001) of Framatome, the French-American constructor of reactors, and Cogema, whose job was to mine uranium and dispose of waste. (See also Commissariat à l'Energie Atomique). At one time, the French relied on domestic uranium, largely from Brittany, but these supplies long ago became impractical and nowadays Areva has vast mining interests across the world, notably in Niger, Kazakhstan, and Canada. This access to world uranium supplies should have been a powerful lever in Areva's efforts to expand sales of its EPR worldwide. Alas, so far, it has struggled to persuade customers that the new rector is worth its hefty price tag. Actually, even in the boom years of nuclear plant construction in France, when the French state ordered large numbers of reactors, it relied on a US design that was licensed from Westinghouse.

**Atom** Just what is an atom anyway? Of course, it is a tiny piece of energy. Every atom has two parts, a nucleus consisting of a certain number of protons and neutrons, surrounded by electrons. In the 19th century, a Russian monk called Mendeleev classified all the atoms present in nature by size. Hydrogen (number 1 in the Periodic Table) is the smallest and hence lightest, while the largest and heaviest is Uranium (at number 92). It is because uranium is so large that it is relatively unstable.

**CFCs (chlorofluorocarbons)** The worldwide ban on the use of CFCs, those anonymous chemicals used, for example, in refrigerators, but later held to be responsible for damage to the ozone layer, is often considered to be an example of how environmental interests can trump economic calculations, at least given the backing of the UN. Cynics point out that the ban actually was in the economic interests of the United States, a country that dominates the United Nations' structures, as US firms held the patents on replacement chemicals.

**Chernobyl** Chernobyl (like Fukushima) is a place whose name has taken on a new and unwanted second meaning. Originally a small city (before its evacuation it had 14,000 inhabitants) in the Ukraine, it is now almost completely abandoned. Ever since its Soviet-era nuclear reactor overheated and exploded, spreading a cloud of radiation across much of western Europe, the consequences of which are still hotly disputed, the city's name became synonymous with nuclear disaster.

**Commissariat à l'Energie Atomique (CEA)** Created at the end of World War II, the CEA had for years almost unlimited funds and operated with the secrecy of a body responsible for providing France with its nuclear weapons. Even today, it is often referred to in France as the "state within the state." In 1976, its industrial activities were split off into a separate business called Cogema, which then merged with Framatome. (See also "Areva.") In recent years, though, it has been more active in the marketplace, not only trying to promote sales of French nuclear expertise, but also investing in renewable energy, such as wind turbines and solar panels—and oddities such as nanotechnology. In 2010, the CEA had a budget of €4 billion and employed no fewer than 16,000 people.

**Depleted uranium (DU)** DU is one of the strange and invariably dangerous by-products, alongside new elements like Plutonium, of running a nuclear reactor to make electricity.

**E.ON** E.ON is a major Germany-based energy company with interests in all kinds of electricity generation, not just nuclear. In recent years it has tried to position itself as "carbon friendly" and after Fukushima has distanced itself from nuclear generation.

**European Bank for Reconstruction and Development (or EBRD)** The EBRD has taken a lead role in clearing up after the Chernobyl disaster, a task so expensive that it dwarfs the ability of national governments acting in isolation.

**Électricité de France (EdF)** EdF is the biggest nuclear generator in the world, with its fleet of French reactors assured plum place in the French market. Created in 1946, as part of a wave of massive nationalization following the end of the Second World War, it was privatized only in 2005, and even then the French state held on to 85 percent of the shares. The great bulk of its electricity is generated via nuclear plants, all provided to it by the French corporation Areva, as part of what the French call a "ménage à trois"—the French government is third partner in the marriage. With a turnover of €66 billion a year, EdF employs 160,000 people but also has large debts. As of 2010, at least €42 billion.

**EPR** The European Pressurized Reactor also sold as the Evolutionary Power Reactor, still keeping the same acronym. Although marketed as the very latest technology, in industry parlance, "Third Generation," the EPR is basically a very large standard pressurized water reactor to which some extra security features have been introduced. Two prototypes are in the course of construction at the time of writing (in 2011)—one in Finland and one in France. Both are behind schedule and the cost of each reactor has been continually reevaluated. They are now supposed to cost about $6 billion each.

**Fukushima** Fukushima (like Chernobyl) is a place as well as a power plant—in its case it is a city and a province (prefecture) on the coast of Japan. The Fukushima Daiichi Nuclear Power Plant itself, though, consists of not one but six reactors along with associated "cooling ponds" in which highly radioactive spent fuel has to be kept for many years. In the aftermath of the earthquake of March 2011, and the tsunami that followed, the outer housings of two of the six reactors exploded, followed by a partial meltdown, fires, and the release of large amounts of radiation.

**International Atomic Energy Agency (IAEA)** Eisenhower proposed an international atomic agency in his 1953 "Atoms for Peace" speech—it came into being four years later in Vienna. A specialized agency within the United Nations framework, its role is to provide advice and technical assistance to countries committing themselves to the peaceful uses of the atom. In this role (which includes safety matters) it has no authority over national governments. However, the UN has also conferred on the IAEA a global policing duty to prevent states seeking to develop the technology for military purposes, backed by all the authority the UN Security Council cares to give it.

**International Energy Agency (IEA)** The IEA publishes the World Energy Outlook reports each year. The body is openly pronuclear, and even its first post-Fukushima report steadfastly insisted that nuclear energy was set to expand by 70 percent in the next 25 years.

**International Panel on Climate Change (IPCC)** The IPCC is a UN body tasked with advising national governments on the latest scientific opinions on possible effects of human activities on the climate, notably the supposed "warming effect" of large-scale releases of carbon dioxide resulting from the burning of fossil fuels. Key figures in the IPCC have always had a friendly eye for nuclear power. The panel was first established in 1988 by the WMO (or World Meteorological Organization) and the UNEP (United Nations Environment Program), both bodies within the UN structure. Despite often being considered the final authority on climate science, the IPCC does not carry out its own original research, nor does it do the work of monitoring climate or related phenomena itself. In recent years, its findings (such as that the Himalayas were set to melt by 2035) have been discredited by revelations that more than a quarter of its supposedly impeccable scientific sources were pressure groups, such as Greenpeace and the Worldwide Fund for Nature.

**Isotopes** When uranium is bombarded with atomic particles, new isotopes (reflecting changes in the cores of its constituent atoms) can be formed with new and different properties. Some of these isotopes are highly unstable, giving off even more atomic particles, and it is this chain reaction that makes possible both nuclear power and nuclear weapons.

**Israel Atomic Energy Commission (IAEC)** The Israeli government body responsible for the management of Israel's "research" nuclear reactors and semi-secret nuclear weapons program. The Israeli prime minister is the chairman of the Atomic Energy Commission.

**Kilowatt (kW)** One kilowatt of power is approximately equal to 1.34 horsepower. One kilowatt hour (kWh) is a unit of energy equal to 1,000 watt hours or 3.6 megajoules. One megawatt (MW) is a million watts, or a thousand kilowatts. One gigawatt (GW) is a billion watts, that is, one thousand megawatts. Last but not least, one terawatt (TW) is a thousand gigawatts.

**Magnox** Magnox is a now obsolete type of nuclear reactor that was originally designed in the United Kingdom (where it is still used). The name comes from the materials used to clad the fuel rods inside the reactor. Over the years the technology has been exported to other countries, where it has been favored not only as a power plant, but as an effective machine for the production of plutonium for bombs.

**MOX** Mixed Oxide fuel (MOX) is cheaper than the uranium fuel used in most reactors but it is especially risky. Created by mixing up plutonium and uranium oxides, it is highly irradiated and hence harder to use, store, and cope with in the event of an accident.

**Nuclear Energy Agency (NEA)** The NEA was originally formed on February 1, 1958, under the name the European Nuclear Energy Agency (ENEA), with the United States participating as an Associate Member. It is an intergovernmental agency that is organized under the OECD, and its particular mission is to "assist its Member countries in maintaining and further developing, through international co-operation, the scientific, technological, and legal bases required for the safe, environmentally friendly and economical use of nuclear energy for peaceful purposes." The prefix "European" was dropped after Japan became a member.

**Nuclear Non-Proliferation Treaty (NPT)** The NPT is a landmark international treaty whose stated objective is to prevent the spread of nuclear weapons and weapons technology and to promote cooperation in the peaceful uses of nuclear energy. Added to this, but not much acted upon, is a goal of achieving nuclear disarmament. Indeed, the Treaty commits its members to the unlikely task of achieving worldwide and complete disarmament. Since it was "opened for signature" in 1968, a total of 190 parties have joined the Treaty, including the five nuclear-weapon states. In fact, more countries have ratified the NPT than any other arms limitation and disarmament agreement. The NPT is often seen to be based on a bargain: In return for eschewing nuclear weapons, states are provided with access to the very latest nuclear technologies.

**Olkiluoto** If you want a nuclear energy holiday, why not visit here? Olkiluoto is an island off the coast of Finland that is home to the Olkiluoto Nuclear Power Plant, consisting of two Boiling Water Reactors and one "under construction" EPR. The island is also the proposed final repository for Finland's nuclear waste.

**Organization for Economic Cooperation and Development (OECD)** The OECD is a grouping of 34 countries that came together in 1961 "to stimulate economic progress and world trade." Its headquarters are in Paris and its membership consists essentially of European and North American nations, plus Japan and Australia. It is generally seen as a "rich man's club."

**Plutonium** Plutonium does not occur naturally; it is a by-product of nuclear reactions. When uranium-238 is persuaded to self-destruct in a nuclear reactor, among the by-products will be three kinds of plutonium, (239, 241, and 242), indicating the number of neutrons in the element's core. It is plutonium-239 that is particularly prized by atomic bomb builders, and each industry standard 900 MW reactor produces about 200 kilos of plutonium each year. Plutonium is highly toxic, and the inhalation of 30 micrograms has been calculated to be enough to kill an adult (by causing cancer). The large stock of it generated by the nuclear industry is thought to pose a particular hazard should terrorists obtain even small amounts of it.

**Pressurized water reactors (PWRs)** PWRs make up the majority of the western nuclear reactor "fleet," as it is known in industry parlance. Indeed, all France's prodigious output of nuclear electricity at the moment comes this way. Originally, however, the technology was developed merely to power submarines.

**Radioactivity (Curies, sieverts, and becquerels)** Radioactivity is a measure of the number of disintegrations produced each second by an unstable element, such as radium or uranium-238. The original unit of measurement of radioactivity was the curie (in honor of the researchers, Marie and Pierre), but since 1986, researchers have instead used a more confusing measure known as the becquerel. One curie corresponds to the activity of one gram of radium-226 per second. It is also equivalent to 37 billion becquerels. The accident at Three Mile Island released 15 curies of radioactive iodine-131 into the environment, whereas Chernobyl released no less than 50 million curies! The sievert (Sv) is a related measurement that better describes the typical annual dose of radioactivity a person receives each year—this is two millisieverts (mSv). It is thought that anything above 100 mSv is dangerous (the effects are largely causing cancer) while a dose of 5000 mSv is reckoned to kill half the people exposed to it within one month.

**Tennessee Valley Authority (TVA)** The TVA is a US electricity company that is 100 percent owned by the US government in Washington. As a result, it has easy access to loans and no concerns about its credit rating. As the nuclear economist Steve Thomas says: "It is therefore not a coincidence that [the TVA] has been at the forefront of efforts to restart nuclear ordering."

**Teollisuuden Voima (TVO)** TVO is the Finnish electric utility and it runs two nuclear reactors. TVO is owned by the major Finnish industrial and power companies, which it then supplies electricity to on a not-for-profit basis. The arrangement means that TVO has a guaranteed market—even if its electricity becomes very expensive.

**TEPCO** The Tokyo Electric Power Company is Japan's major electricity utility. Among its nuclear plants is the Fukushima complex. TEPCO's liability insurance was very small, minuscule even relative to the damages caused. By late April 2011, the costs and economic damage from the Fukushima disaster were already being estimated in the range of $130–175 billion, of course rising almost by the day. By comparison, TEPCO's consolidated total turnover for its last full year before the disaster, 2010, was about $57 billion, which generated net operating profits of about $3.15 billion. This indicates that a good 50 years of the company's total operating profits would be needed to cover even the early estimates of the total economic damage caused. By the end of 2011, with its share price languishing at 10 percent of its pre-crisis days, the Japanese government was openly talking with the company about "nationalization."

**UNSCEAR** The United Nations Scientific Committee on the Effects of Atomic Radiation was set up in 1955. Its 21 participating member countries provide scientists to serve as members of a committee that holds formal meetings (sessions) annually and submits a report to the General Assembly. Like all UN bodies, it is highly political and its generally pronuclear findings should be treated with caution.

**Uranium** Naturally occurring uranium comes in essentially two flavors, distinguished by the number of subatomic particles in its nucleus. The most common kind of uranium, some 99.3 percent, is uranium-238, which is hardly radioactive at all. The remaining 0.7 percent of the element (as it occurs in nature) consists of uranium-235. Uranium-235's nucleus (consisting of 143 neutrons and 92 protons) is inherently unstable and under certain conditions can be persuaded to "split," releasing the much sought-after energy. To extract the useful, radioactive uranium from the useless stable kind requires in itself a great deal of effort, energy, and cost. In the process, too, large quantities of radioactive waste are produced, including new "unnatural" elements, such as plutonium. To make fuel for a nuclear reactor, the proportion of the radioactive uranium-238 isotope in the mineral must be raised to about 4 percent; to make an atomic bomb, however, the proportion must be raised to near 90 percent.

**Westinghouse** Westinghouse is a venerable US electricity company (established in 1886) that became the US government's furnisher of nuclear reactors after the end of World War II. Most of the world's nuclear reactors use its technology. These days it is owned and operated by the Toshiba Group but it is still based in Pennsylvania in the United States.

**World Health Organization (WHO)** The WHO is charged with promoting health worldwide, but it has an extra duty as regards nuclear. As part of an agreement signed on May 28, 1959, the IAEA has supervision of all its policies regarding nuclear and related issues, so all WHO reports on nuclear accidents like that of Chernobyl have effectively been subcontracted to the IAEA.

**Yucca Mountain** For many years, Yucca Mountain, an escarpment in Nevada near the border with California, was for many years supposed to provide a final resting place for all the radioactive waste created by the US nuclear energy program, but in 2010, following a court case by energy companies themselves (tired of paying for its development), it was finally admitted that the repository was never going to be built.

# NOTES, KEY SOURCES, AND SUGGESTIONS FOR FURTHER READING

Where documents are available in electronic form on the web, these links have been preferred and page references are not given. Similarly, rather than misdirect readers with often cumbersome links that may in any case now be moved or broken, sometimes we have preferred to provide key word information for electronic searches of the web, for example, using Google.

## FOREWORD

p. xii The UK government report "whose main recommendation was that there should be no commitment to a large program of nuclear fission power" was the so-called "Flowers Report" by Sir Brian Flowers, "Sixth Report UK Royal Commission on Environmental Pollution," 1976.

## INTRODUCTION AND OVERVIEW

p. 4 The Queen's speech was reported in full in the *West Cumberland News*'s celebratory front page story on October 17, 1956.

p. 8 "Each new reactor will produce as much as 50 kilograms of plutonium . . . a year": See, for example, *Reviews of Modern Physics, Vol. 50, No. 1, Part II* (January 1978) or the more general discussion of the Nuclear Information and Resource Service at http://www.nirs.org/factsheets/plutbomb.htm.

p. 9 Marie Curie and the early research into radioactivity is described in a clear and readable form in the *Nobel Lectures, Physics 1901-1921* (Amsterdam: Elsevier, 1967).

pp. 10, 15 The estimates of the number of atomic weapons created outside of the the NPT are the authors' own estimates, drawing on widely available published information, such as that by The International Atomic Energy Agency (IAEA). The IAEA is an international organization (based in Vienna, Austria) specifically set up in 1957 to promote what it calls "the peaceful use of nuclear energy." Though established independently of the United Nations through its own international treaty, the IAEA Statute, the IAEA reports to both the UN General Assembly and Security Council. Its website, www.iaea.org, contains much factual material and research resources.

p. 17 "An initial fuel load of around 150 tons of plutonium": See, for example, *Nuclear Energy Outlook* (OECD Nuclear Energy Agency, 2008).

p. 17 A discussion of the different kinds of nuclear reactors and their toxic appetites is available at http://www.3rd1000.com/nuclear/nuke101g.htm.

pp. 18-19 The pride of Indian politics is described in "Xii Lok Sabha Debates, Session II, (Budget)," www.fas.org/news/india/1998/05/0829059801.htm.

p. 19 Transparency International (www.transparency.org) is the global civil society organization leading the fight against corruption. It was originally founded in Germany and is headquartered in Berlin.

## MYTH 1: NUCLEAR ENERGY IS THE ENERGY OF THE FUTURE

This chapter is essentially an overview of the ideas and issues that are explored in more detail later in the book. For those who wish to follow up particular examples in this chapter however, the following details are offered:

pp. 31-33 The figure of 16 percent for the contribution of nuclear power in France to the country's total energy needs comes from Mycle Schneider, "Nuclear Power in France—Beyond the Myth," commissioned by the Greens-EFA Group in the European Parliament, Brussels, December 2008.

The sources for the Amazing Fact are: M. Schneider, A. Froggatt, and S. Thomas, "World Nuclear Industry Status Report, 2010-2011," and the authors' own original research.

p. 34 The estimate of the precise prices "needed" to make nuclear energy economic, but not these opinions, were given by Martin Young, head of European Utilities Equity Research, Nomura, London, in "The Nuclear Renaissance: An Equity Analyst's Perspective," a paper presented to the 35th Annual Symposium of the World Nuclear Association, September 2010.

p. 38 The "green plea" appears in Ron Cameron and Martin Taylor, "The 2050 Roadmap for Nuclear: Making a Global Difference," *Energy & Environment* 22, nos. 1 & 2 (2010). The paper was originally presented to the 35th Annual Symposium of the World Nuclear Association in September 2010.

*Other data sources and suggestions for further reading for this chapter:*

In writing this chapter, we have benefited from the clear overview in *The Economics of Nuclear Power,* Nuclear Issues Paper No. 5 (2005) and *The Economics of Nuclear Power: An Update* (March 2010), both written by Steve Thomas and published by the Heinrich Böll Foundation.

*Other relevant perspectives on the issue include:*

"Energy Policies of IEA Countries—2003 Review," published by the Organization for Economic Cooperation and Development/International Energy Agency.

The European Investment Bank, "Energy Review," October 2006.

"The Future of Nuclear Power," Massachusetts Institute of Technology, 2003, www. web.mit.edu/nuclearpower/.

"The Future Role of Nuclear Energy in Europe," World Energy Council, Alessandro Clerici (chair), June 13, 2006.

The Paris-based International Energy Agency runs a website, www.iea.org, which includes many resources, such as energy supply statistics and projections.

"Outlook on Advanced Reactors," *Nucleonics Week* Special Report, March 30, 1989.

## MYTH 2: NUCLEAR POWER IS GREEN

This chapter sketches the wider political context of recent energy debates and draws on two long articles Martin Cohen wrote for the *Times Higher Education* (London): "Beyond Debate," December 9, 2009, and "Profits of Doom," July 22, 2010, both of which are available online at www.

timeshigher.co.uk, albeit without the rather fine pictures in the original magazine. The original articles contain more detail, but there is also a vigorous debate online and some additional ideas in the numerous comments that follow the articles. Climate science is evidently a topic that many people have a view on.

We also consulted numerous specialist papers, such as the International Panel on Climate Change reports, especially the "Summaries for Policy Makers"; papers in various obscure scientific journals, such as one by "warmists" Gavin Schmidt, Drew Shindell, Ron Miller, Michael Mann, and David Rind, "General Circulation Modelling of Holocene That Is, the Current Climate Period Climate Variability," *Quaternary Science Review* 23 (2004); and the fence-sitters Richard S. Lindzen and Yong-Sang Choiand, "On the Determination of Climate Feedbacks from ERBE Data." We also examined numerous contrarian reports by climate skeptics, often found only on the Internet. However, if you want a good objective assessment of climate science, a book that came out after the Copenhagen conference by Robert Carter, *Climate: The Counter Consensus* (London: Stacey International, 2010), tells you all you need to know.

The scholarly work, *International Environmental Policy,* by Aynsley Kellow and Sonja Boehmer-Christiansen (Edward Elgar, 2002) provides an authoritative insight into the political forces driving climate research. A companion work, *The International Politics of Climate Change* (2010), is a good resource for libraries—it costs hundreds of zlotys! A pity, as it has a good look at what it calls "The Place of Science."

*On a few of the specific quotes in the chapter:*

p. 40 John Ritch, director general of the World Nuclear Association, put it like that in a speech in June 2007.

p. 43 More pronuclear pleas are from *Nuclear Lessons Learned* (London: Royal Academy of Engineering on behalf of Engineering the Future, October 2010) with main research by Malcom Joyce, Richard Garnsley, and Ian Nickson, available at www.raeng.org.uk/news/publications/list/reports/Nuclear_Lessons_Learned_Oct10.pdf.

p. 48 Timothy Wirth's infamous promise to "ride this global warming thing" is recorded alongside other revealing quotes at the Science and Public Policy Institute website, http://scienceandpublicpolicy.org/policy.html.

p. 49 The "Second Assessment Report" of the IPCC is available at www.ipcc.ch/pdf/ . . . /ipcc . . . assessment/2nd-assessment-en.pdf.

p. 52 "Why Greens Must Learn to Love Nuclear Power," by Mark Lynas, appeared in *New Statesman,* September 18, 2008, and is reproduced online at http://www.newstatesman.com/environment/2008/09/nuclear-power-lynas-reactors.

Lovelock's supporting view, as published in the May 24, 2004 issue of the *London Independent* newspaper (Opinion; Pg. A13) under the headline "Nuclear Power is the Only Green Solution," is reproduced appreciatively at http://www.nuclear.com/archive/2004/05/31/tbd and at http://www.jameslovelock.org/page11.html.

pp. 53-55 James Lovelock recommends nuclear salvation and describes "The Revenge of Gaia" in a book of the same name (New York: Penguin, 2007).

p. 56 The Venus article was in *Popular Science* (June 1982), pp. 52-58.

pp. 57-58 Bruce Yandle, an economist at Clemson University in the United States, coined the phrase "Baptist and Bootlegger coalitions" in an article in *Regulation Magazine* in 1983.

p. 61 The "drowning bears" photostory was disowned by Denis Simard in the *National Post,* http://www.nationalpost.com/story.html?id=5961259b-de08-4532-850b-09d4753bed39&k=88988.

p. 62 "European imports of coal in 2009 were over 300 million tons . . .": The interesting truth about Europe's need for coal is given in a chart here: http://www.indexmundi.com/energy.aspx?region=eu&product=coal&graph=imports.

*Other data sources and suggestions for further reading for this chapter are:*

John Holdren's warning is in Paul Ehrlich's (one of global warming theory's most extravagant doomsayers) book, *The Machinery of Nature* (New York: Simon & Schuster, 1986), p. 274.

Organization for Economic Cooperation and Development/International Energy Agency, *World Energy Outlook* (annual reports).

Organization for Economic Cooperation and Development/International Energy Agency, "Renewable Energy Policy in IEA Countries," 1998.

## MYTH 3: NUCLEAR REACTORS ARE RELIABLE AND SAFE

This chapter also draws on the two articles Martin Cohen wrote for the *Times Higher Education* (London): "Beyond Debate," December 9, 2009, and "Profits of Doom," July 22, 2010.

p. 66 The source for details on the Windscale Fire is the British Nuclear Fuels, PLC official Company History, available at http://www.fundinguniverse.com/company-histories/BRITISH-NUCLEAR-FUELS-PLC-company-History.html. At the time, the AEA rushed to assure an anxious public that radiation levels were still only a tenth of that considered dangerous. It was not until the 1980s that a report was released suggesting that the level of radiation near Windscale after the accident was up to 40 times greater than had originally been claimed.

p. 69 The Gundersen quote on the true scale of the problems at Fukushima is reported in the *Huffington Post:* http://huffingtonpostunionofbloggers.org/2011/06/18/more-nukes-japan-risks-becoming-totally-uninhabitable.

p. 75 The Japan Nuclear Energy Safety Organization consists of various expert engineer groups joining together to "ensure the nuclear safety." It runs a website at www.jnes.go.jp/english/index.html.

pp. 71, 82 An online article by Steven Swinford and Christopher Hope,"Japan Earthquake: Japan Warned Over Nuclear Plants, WikiLeaks Cables Show," March 15, 2011, gives more details of how the Japanese regulator has historically minimized the need to protect plants against earthquakes: http://www.telegraph.co.uk/news/worldnews/wikileaks/8384059/Japan-earthquake-Japan-warned-over-nuclear-plants-WikiLeaks-cables-show.html.

p. 72 Mr. Bajaj says that despite being attached to the Atomic Energy Commission, his agency is "functionally independent" of the country's nuclear establishment in an interview given to the *New York Times,* "Resistance to Jaitapur Nuclear Plant Grows in India," April 13, 2011, http://www.nytimes.com/2011/04/15/business/global/15nuke.html.

p. 73 Guillaume Wack's reassurances on the French nuclear situation are revealed again in the *New York Times* in an article, "Safety Fears Raised at French Reactor," July 26, 2010, http://www.nytimes.com/2010/07/27/business/global/27iht-renepr.html.

p. 74 The words of former governor of Fukushima provinces, Eisaku Satō, were widely reported at the time. See for example, the article "Regulators Ignored Warning Signs at Fukushima By Evann Gastaldo," March 22, 2011, http://www.newser.com/story/114676/japan-nuclear-crisis-at-fukushima-dai-ichi-nuclear-plant-regulators-granted-extension-despite.html.

p. 75 Tokoo Kano's cozy relationship with the Japanese government is described by the *New York Times* in an interesting special investigation by Norimitsu Onishi and Ken Belson, "Culture of Complicity Tied to Stricken Nuclear Plant," April 26, 2011, http://www.nytimes.com/2011/04/27/world/asia/27collusion.html.

pp. 75-76 "Chernobyl: Consequences of the Catastrophe for People and the Environment," a translation of a 2007 Russian publication by Alexey V. Yablokov, Vassily B. Nesterenko, and Alexey V. Nesterenko, was published by the New York Academy of Sciences in 2009 in their Annals of the New York Academy of Sciences series. It is at http://www.strahlentelex.de/Yablokov%20Chernobyl%20book.pdf.

pp. 79-81 *Voices from Chernobyl: The Oral History of a Nuclear Disaster* by Svetlana Alexievich, is available in a translation by Keith Gessen published by Picador. The text is also widely available online, e.g., http://www.faylicity.com/book/book1/fstchernob.html.

p. 83 The daring exploits of the plant workers at Fukushima are described in a widely syndicated *New York Times* article by Keith Bradsher and Hiroko Tabuch, "Last Defense at Troubled Reactors: 50 Japanese Workers," March 15, 2011, http://www.nytimes.com/2011/03/16/world/asia/16workers.html.

*Other sources consulted include:*

European Bank for Reconstruction and Development, "Shelter Implementation Plan, Chernobyl Shelter Fund," February 2000, www.iaea.org/newscenter/features/chernobyl-15/shelter-fund.pdf.

S. Hirschberg, G. Spiekerman, and R. Dones, "Severe Accidents in the Energy Sector," Paul Scherrer Institut for the Swiss Federal Office of Energy, November 1998; http://manhaz.cyf.gov.pl/manhaz/szkola/materials/S3/psi_materials/ENSAD98.pdf.

International Atomic Energy Agency (IAEA), "Chernobyl Accident: Updating of INSAG-1 Safety Series," No.75-INSAG-7 (Vienna, 1991), p. 24.

IEAE, "Ten Years after Chernobyl: What Do We Really Know?" April 1996. The figure of 400 times for the radiation released from Chernobyl is the IAEA's own one, from its website FAQ page on Chernobyl.

*NEI Source Book: Fourth Edition* (NEISB_3.3.A1), www.insc.anl.gov/neisb/neisb4/NEISB_3.3.A1.html.

"Nuclear Lessons Learned" (London: Royal Academy of Engineering on behalf of Engineering the Future, October 2010); with main research by Malcom Joyce, Richard Garnsley and Ian Nickson, available at www.raeng.org.uk/news/publications/list/reports/Nuclear_Lessons_Learned_Oct10.pdf.

World Health Organization, "Health and Environment in Sustainable Development: Five Years after the Earth Summit" (Geneva, Switzerland: Author, June 1997); http://whqlibdoc.who.int/hq/1997/WHO_EHG_97.12_eng.pdf.

## MYTH 4: NUCLEAR ENERGY IS "TOO CHEAP TO METER"

In this chapter, focusing on the economics again, we have particularly benefited from the groundbreaking research of Steve Thomas in a paper for the Heinrich Böll Foundation in 2005: *The Economics of Nuclear Power,* Nuclear Issues Paper No. 5. It was published, as the Foundation put it, rather optimistically, as a contribution to the debates on the future of nuclear energy intended to coincide with the twentieth anniversary of the Chernobyl disaster. Another comprehensive overview is Henry Sokolski, ed., *Nuclear Power's Global Expansion: Weighing Its Costs and Risks* (Strategic Studies Institute, December 2010), although not necessarily expressing the views of the SSI, which is part of the US Army War College and based in Pennsylvania.

p. 90 Details of the troubled Olkiluoto project from the sympathetic perspective of the energy industry itself are at http://www.power-technology.com/projects/Olkiluoto/.

The clouds gathering over the French nuclear industry are revealed in various Bloomberg news reports including: "Areva Risks Downgrade Amid Limited Capital Increase," December 14, 2010, http://www.businessweek.com/news/2010-12-14/areva-risks-downgrade-amid-limited-capital-increase.html, and the very informative booklet (in French), "Nucleaire: C'est par où la sortie?" (*Les dossiers du Canard enchaine* (2011), pp. 89, 91.

The implications of the French nuclear industry's debts in the slightly less dire situation of 2008 are discussed in the Public Services International Research Unit report by Steve Thomas, "Areva and EDF: Business prospects and risks in nuclear energy," March 2009.

p. 93 The Nuclear Energy Agency report that optimistically assessed the prospects for bulk savings and the comments of GEH boss John Fuller ("GEH: Cost estimates did industry a disservice," September 17, 2009) are both quoted in Steve Thomas' "The Economics of Nuclear Power: An Update," March 2010, www.boell.de/downloads/ecology/Thomas_economics.pdf.

p. 94 The figures about the financial clouds gathering over Électricité de France and Areva are drawn from correspondence between the authors and John Busby, July 29, 2010. The closing quotation appeared in the *Economist,* May 19, 2001, pp. 24-26.

pp. 96, 99 *Nucleonics Week* itself is available online at http://www.platts.com/Products/nucleonicsweek but only to subscribers.

The 2009 annual report for EDF is at http://www.edf.com/html/RA2009/uk/index.html and Areva offers details of its finances at http://www.areva.com/EN/finance-57/uptodate-transparent-economic-information-for-the-financial-community.html.

pp. 97, 99 "Nuclear Loan Guarantees: Another Taxpayer Bailout Ahead?" is at http://www.ucsusa.org/nuclear_power/nuclear_power_and_global_warming/nuclear-loan-guarantees.html.

p. 99 The power chief's impassioned plea for subsidies is described in an article by Rowena Mason for the *Daily Telegraph* (London), "Britain's Power Chiefs Reveal Nuclear Blueprint," November 13, 2010, http://www.telegraph.co.uk/finance/newsbysector/energy/8131700/Britains-Power-chiefs-reveal-nuclear-blueprint.html.

pp. 99-100 A discussion of energy efficiency and capacity factors is at http://www.mpoweruk.com/energy_efficiency.htm.

p. 105 Henry Sokolski, ed., *Nuclear Power's Global Expansion: Weighing Its Costs and Risks* (Strategic Studies Institute, December 2010).

p. 113 "The Stern Review on the Economics of Climate Change" is available at http:/www.hm-treasury.gov.uk/sternreview_index.htm.

pp. 113-114 The Union of Concerned Scientists report is at: http://webarchive.nationalarchives.gov.uk/+/.

*Other data sources and suggestions for further reading for this chapter are:*

Eurelectric, "A Quantitative Assessment of Direct Support Schemes for Renewables," 2004.

European Environment Agency. "Energy Subsidies in the EU," *EC Inform-Energy*, June 2004.

ExternE web site, EC Inform—Energy #96, 9/01, EU DG Research 2003, External costs, on ExternE website.

Malcolm Grimston and Peter Beck, "Nuclear Energy Research, Development and Commercialisation," RIIA (draft), 2001.

House of Commons Energy Select Committee, "Fourth Report: The Costs of Nuclear Power," HMSO, June 1990.

House of Lords Science & Technology Committee, "Renewable Energy: Practicalities," 2004.

International Energy Agency, "Projected Costs of Generating Electricity—2005 Update" (Paris: 2005).

Krewitt et al., "Environmental Damage Costs from Fossil Electricity Generation in Germany and Europe," *Energy Policy* 27 (1999): 173-183.

A. McDonald and L. Schrattenholzer, "Learning Rates for Energy Technologies," *Energy Policy* 29 (2001): 255-261.

PB Power, "Powering the Nation: A Review of the Costs of Generating Electricity" (Newcastle, UK: Author, 2006), www.pbpower.net/inprint/pbpubs/powering_the_nation_summary.pdf.

Performance and Innovation Unit, "Energy Review Working Paper: The Economics of Nuclear Power," 2002.

Royal Academy of Engineering, "The Costs of Generating Electricity" (London: 2004).

US Department of Energy, "An Analysis of Nuclear Power Construction Costs, Energy Information," DOE/EIA-0411, 1986.

US National Research Council, "Hidden Costs of Energy: Unpriced Consequences of Energy Production and Use," 2009.

## MYTH 5: NUCLEAR POWER TRUMPS GEOPOLITICS

pp. 127-137 A good part of the core directions for this "Whistle-stop Tour" of the world is drawn from academic studies by the Public Services International Research Unit at Greenwich University in the United Kingdom. Additional research and information is from correspondence between the authors and nuclear policy analyst John Busby, especially email, July 29, 2010. The quote on the Labour government in the United Kingdom's changing policy is from "Labour Prepares to Tear Up 12 Years of Energy Policy," *The Times*, February 1, 2010.

*But now some extra guidance on the "Whistle-stop Tour":*

The United States-Vietnam negotiations were widely reported at the time, for example, see the article "U.S., Vietnam in civilian nuclear talks," August 5, 2010, http://af.reuters.com/article/idAFN0516727220100805.

"Israel's Nuclear Weapons—The Secrecy Must Stop," February 11, 2009, http://www.opednews.com/articles/Israel-s-Nuclear-Weapons-by-Joe-Parko-100519-274.html. A related and rather fascinating essay by Colonel Warner D. "Rocky" Farr, for the Federation of American Scientists, "The Third Temple's Holy Of Holies: Israel's Nuclear Weapons," http://www.fas.org

/nuke/guide/israel/nuke/farr.htm. And see also Seymour M., Hersh, *The Samson Option. Israel's Nuclear Arsenal and American Foreign Policy* (New York: Random House, 1991), 223.

The detail on Turkey is from Lauri Marchan, "Ankara maintient ses projects nucleaires en zone sismique," *Le Figaro,* March 18, 2011, www.lefigaro.fr/ . . . /01003-20110317ARTFIG00790-ankara-maintient-ses- projets-nucleaires-en-zone-sismique.php.

*Other data sources and suggestions for further reading for this chapter are:*

J. H. Williams and F. Kahrl, "Electricity Reform and Sustainable Development in China," IOP Electronic Journals, 2008, www.iop.org/EJ/article/1748-9326/3/4/044009/erl8_4_044009.html#erl301237s3.2.

Bruce Einhorn, "UN and China Squabble over Wind Subsidies," Bloomberg Businessweek, December 2, 2009; www.businessweek.com/globalbiz/blog/eyeonasia/archives/2009/12/un_and_china_sq.html.

International Energy Agency, "Energy Policies in IEA Countries, Country Review," 2001.

An article by the Federation of American Scientists (citing various books) on the history of the Israeli bomb is at www.fas.org/nuke/guide/israel/nuke/.

## MYTH 6: NUCLEAR ENERGY IS VERY CLEAN

p. 146 For more on the World Nuclear Association's view of radiation hazards, see http://www.world-nuclear.org/info/inf25.html. The quotation used here was accessed December 12, 2011.

p. 152 UPI's report on the Asse nuclear waste site, from Berlin (UPI), was dated September 8, 2008.

pp. 153, 156 Greenpeace has many pamphlets, some of them quite serious, such as "Nuclear Power: A Hazardous Obstacle to Clean Solutions" (2009). All are available as pdfs at their website, www.greenpeace.com.

p. 155 By April 2011, the cost of the Chernobyl Sarcophage was being put at 1.54 billion euros, three times the original estimate of 432 million. See news reports, such as: "Tchernobyl: le nouveau sarcophage a pris . . ." *Le Figaro—International,* April 19, 2011, www.lefigaro.fr/ . . . /01003-20110419ARTFIG00629-tchernobyl-le-nouveau-sarcophage-a-pris-du-retard.php.

p. 156 "EDF to Place French Power Grid into Nuclear Dismantling Fund," agency news reports, July 21, 2010.

p. 158 Jalal Ghazi monitors and translates Arab media for New America Media. For more details on the DU issue, see "Cancer—The Deadly Legacy of the Invasion of Iraq," January 6, 2010, http://news.newamericamedia.org/news/view_article.html?article_id=80e260b3839daf2084fdeb0965ad31ab.

pp. 159-60 That high-tech Texas University strategy is online at www.eurekalert.org/pub_releases/2009-01/uota-nfh012709.php.

pp. 160-61 George Monbiot rattles on in his *Guardian* blog about being "Nuked by Friend and Foe," February 20, 2009, http://www.guardian.co.uk/environment/georgemonbiot/2009/feb/20/george-monbiot-nuclear-climate, repeated at http://www.monbiot.com/2009/02/20/nuked-by-friend-and-foe/.

p. 162 Sandra Upson, "Finland's Nuclear Waste Solution: Scandinavians Are Leading the World in the Disposal of Spent Nuclear Fuel," *IEEE Spectrum,* December 2009. Her optimistic account has to be set against the doom-laden assessment by Harri Lammi, "Nuclear Waste Repository Plans in Finland: Finnish Solution to Nuclear Waste—Silencing the Debate," October 2010. Lammi is program director of Greenpeace Nordic.

*Other data sources and suggestions for further reading for this chapter are:*

Helmut Hirsch, Oda Becker, Mycle Schneider, and Antony Froggatt, "Nuclear Reactor Hazards—Ongoing Dangers of Operating Nuclear Technology in the 21st Century," Greenpeace International (April 2005).

Jackson Consulting, "Paying for Nuclear Clean-Up: An Unofficial Market Guide," (October 2006).

### MYTH 7: NUCLEAR RADIATION IS HARMLESS

This is a controversial area. It is, of course, impossible to prove everything, but here are a few starters for further investigation:

p. 168 Helen Caldicott was opposing nuclear power in a debate called "Prescription_for_Survival" on March 30, 2011.See also the debate at: http://www.democracynow.org/2011/3/30/prescription_for_survival_a_debate_on.

pp. 169-70 The opinions of Lovelock are drawn again from the *Independent,* "Nuclear Power is the Only Green Solution," May 24, 2004.

p. 170 Images of the actual UK government nuclear propaganda memoranda (heavily redacted) are at http://www.guardian.co.uk/environment/interactive/2011/jun/30/email-nuclear-uk-government-fukushima, in a feature "UK Government and Nuclear Industry Email Correspondence After the Fukushima Accident," June 30, 2011. "We do not want to be on the back foot with this. People at new build sites are likely to be following closely," wrote one civil servant.

p. 170 See Max Hastings' sterling defense of nuclear in the wake of the tsunami, "Yes, Nuclear Power Plants Are Dangerous. But for Britain, the Alternative Is to Start Hoarding Candles," *Daily Mail,* March 16, 2011, http://www.dailymail.co.uk/news/article-1366274/Japan-tsunami-earthquake-Nuclear-power-plants-dangerous.html#ixzz1GfAi1U4f.

p. 171 The Calculation of Reactor Accident Consequences) computer program CRAC 2 is at http://www.oecd-nea.org/tools/abstract/detail/ccc-0419/.

p. 172 "Ocean currents will disperse the radiation" reported Reuters in an article by Yoko Kubota, "Engineers Toil to Pump Out Japan Plant," March 26, 2011, http://www.reuters.com/article/2011/03/26/us-japan-quake-idUSTRE72A0SS20110326.

p. 173 The hidden story of nuclear's contract workers is given by a French article at *L'Humanité:* "Pour travailler à Fukushima, il faut être prêt à mourir." Interview by Anne Roy, translated April 7, 2011, by Henry Crapo and reviewed by Bill Scoble.

A summary by Prof. Paul Jobin,"Dying for TEPCO? Fukushima's Nuclear Contract Workers," http://www.globalresearch.ca/index.php?context=va&aid=24543.

See also the *New York Times* feature by Hiroko Tabuchi, "Japanese Workers Braved Radiation for a Temp Job," April 9, 2011, http://www.nytimes.com/2011/04/10/world/asia/10workers.html.

p. 174 Debora McKenzie, "Fukushima Radioactive Fallout Nears Chernobyl Levels," March 24, 2011, http://www.newscientist.com/article/dn20285-fukushima-radioactive-fallout-nears-chernobyl-levels.html.

pp. 175, 183 Alexey Yablokov, Vassily Nesterenko, and Alexey Nesterenko, *Chernobyl: Consequences of the Catastrophe for People and the Environment,* vol. 1181 (New York: Annals of the New York Academy of Sciences, 2009).

pp. 175-76, 177-79 For a bit of light reading but also many more atomic facts, the story of the early atomic tests in the Pacific is in Martin Cohen's political travel guide, *No Holiday: 80 Places You* Don't *Want to Visit* (New York: Disinformation Press 2006).

p. 176 Emilia A. Diomina, "Radiation Epidemiological Studies in a Group of Liquidators of the Chernobyl Accident Consequences," R. E. Kavetsky Institute of Experimental Pathology, Oncology and Radiobiology of National Academy of Sciences of Ukraine, Kiev, 2006.

p. 180 "Ukraine: Radiation haunts children born after the Chernobyl disaster in 1986." For details of the "other side" of the Ukrainian health consequences debate, as collected by environmentalists for a television program, see: http://www.itnsource.com/shotlist//RTV/2006/04/20/RTV580206/?s=Disaster.

p. 181 The *Guardian* website also carries a January 10, 2010 article, "Chernobyl Nuclear Accident: Figures for Deaths and Cancers Still in Dispute," by its environment editor, evidently risking the wrath of his colleague George Monbiot, John Vidal, and another responding to the WHO analysis titled, "UN Accused of Ignoring 500,000 Chernobyl Deaths," March 25, 2006. The influential World Health Organization report is summarized in "Health Effects of the Chernobyl Accident: An Overview," Fact Sheet No. 303, WHO, April 2006.

p. 182 A. E. Okeanov and E. A. Sosnovskaya, "Incidence of Malignant Tumors Among Different Groups of Belarusian Population Affected to the Chernobyl Accident," International State Environmental University, Minsk, Republic of Belarus and Republican Research—Practical Center of Radiation Medicine and Human Ecology, Gomel, Republic of Belarus, 2006.

p. 182 Chris Busby, "Infant Leukaemia in Europe after Chernobyl and Its Significance for Radiation Protection: A Meta-Analysis of Three Countries Including New Data from the United Kingdom," University of Liverpool, Department of Human Anatomy and Cell Biology, and Green Audit, Aberystwyth, UK, 2006.

pp. 182, 189 Yuri Orlov, Andrey Shaversky, and V. Mykhalyuk, "Intracranial Neoplasms in Infants of Ukraine. An Epidemiological Study," Institute of Neurosurgery A.P. Romodanov, AMSU, Kiev, 2006. The authors state specifically: "Deteriorated radiation situation in Ukraine has adversely affected the brain tumor incidence in infants thereby leading to over 2.3 times growth of total patient population and 6.2 times growth in the number of patients under 1 year."

p. 182 Recommendations of the European Committee on Radiation Risk, by C. Busby, R. Bertell, I. Schmitz-Feuerhake, and A. Yablokov, "The Health Effects of Ionising Radiation Exposure at Low Doses for Radiation Protection Purposes," Regulators Edition; ECRR: Brussels, Belgium, 2003.

p. 183 Interestingly, for a journalist who preaches the importance of both consulting and giving your sources, a recent article where George Monbiot declared his newfound love for nuclear power offers as one of his sources the web comic xkcd.com. The comic's author states firmly: "I'm not an expert in radiation and I'm sure I've got a lot of mistakes in here." Bravo for xkcd science! Monbiot's other source, and indeed the mainstream view on the effects of radiation, is the WHO report "Health Effects of the Chernobyl Accident and Special Health Care Programmes," 2006.

p. 183 For more on the Russian researchers' warning that "Chernobyl's radioactive contamination . . . brings the estimated death toll to about 900,000, and that is only for the first 15 years after the Chernobyl catastrophe," see Alexey Yablokov, Vassily Nesterenko, Alexey Nesterenko, *Chernobyl Consequences of the Catastrophe for People and the Environment,* vol. 1181 (New York Academy of Sciences, 2009).

pp. 183-84, 187 Dr. Windridge's equally fine broadside for nuclear appeared in the Guardian, "Fear of Nuclear Power Is Out of All Proportion to the Actual Risks," April 4, 2011, http://www.guardian.co.uk/science/blog/2011/apr/04/fear-nuclear-power-fukushima-risks.

p. 185 *The New Scientist* has covered Fukushima extensively. See http://www.newscientist.com/article/dn20285-fukushima-radioactive-fallout-nears-chernobyl-levels.html.

pp. 185, 186-87 A disquieting paper by B. A. Muggenburg et al., "Toxicity of Inhaled Plutonium Dioxide in Beagle Dogs," *Journal Radiation Research* vol. 145, no. 3 (1996): 361-381, http://www.mendeley.com/research/toxicity-inhaled-plutonium-dioxide-beagle-dogs-1/.

p. 188 "The WHO expert group placed particular emphasis on scientific quality," said WHO Fact Sheet No. 303, "Health Effects of the Chernobyl Accident: An Overview," published in April 2006.

p. 190 "At least 500,000 people—perhaps more—have already died out of the 2 million people who were officially classed as victims of Chernobyl [nuclear power station] in Ukraine": Nikolai Omelyanets, deputy head of the National Commission for Radiation Protection in Ukraine, *The Guardian,* March 25, 2006.

p. 190 Shocking photographs, taken by Paul Fusco, of an orphanage in Minsk filled with children deformed by Chernobyl can be found at http://inmotion.magnumphotos.com/essay/chernobyl.

pp. 192-93 Tony Benn has his own page on nuclear and what's dodgy about it: http://www.tonybenn.com/nucl.html.

p. 193 For the history of bombing, see Sven Lindqvist, *A History of Bombing,* trans. Linda Haverty Rugg (New Press, 2003).

p. 193 The *New York Times/International Herald Tribune* has covered the Fukushima crisis in depth, and several of the stories here giving insights into the aftermath can be traced back to these sources. For example, on August 9, 2011, the *Tribune* gave a front-page scoop on how official computer maps generated by the System for the Prediction of Environmental Emergency Dose Information of likely radiation contamination had been kept secret from the Japanese public.

*Other data sources and suggestions for further reading for this chapter are:*

The book *Nuclear Servitude: Subcontracting and Health in the French Civil Nuclear Industry* (Work, Health and Environment Series) by Annie Thebaud-Mony (Baywood Publishing Company, 2010) caused something of a stir, detailing as it does that the French nuclear industry relies on staff who are paid barely above the French minimum wage (people who are so poor that they even sleep rough near the nuclear plants), to carry out crucial maintenance jobs.

United Nations Scientific Committee on the Effects of Atomic Radiation (UNSCEAR), 2000. "Sources and Effects of Ionizing Radiation," Report to the General Assembly, "Annex J: Exposures and Effects of the Chernobyl Accident," pp. 451-566, 2000.

UNSCEAR, "Health Effects Due to Radiation from the Chernobyl Accident," Draft report A/AC.82/R.673, 2008, pp. 1-220.

## MYTH 8: EVERYONE WANTS TO INVEST IN NUCLEAR ENERGY

p. 199 " . . . it's almost unfinanceable in today's environment": See "Peak Oil News: A Compilation of New Developments, Analysis, and Web Postings," November 18, 2010.

p. 200, 201 The US Energy Information Administration freely offers all its musings on nuclear at http://www.eia.gov/.

p. 200, 203 The Congressional Research Service estimates of the taxpayer liabilities for U.S. nuclear are summarized in "The Economics of Nuclear Power: An Update," by Steve Thomas, March 2010, p. 34, from Congressional Budget Office, *Cost estimate of S.14, Energy Policy Act of 2003* (Washington, DC: Congressional Budget Office, May 7, 2003), http://www.cbo.gov/doc.cfm?index=4206.

p. 201 "Standard & Poor's and Moody's would have a heart attack. And my chief financial officer would, too": *Beyond Nuclear Fact Sheet,* www.psr.org/nuclear-bailout/resources/nuclear-powers-toxic-assets.pdf.

p. 202 Those alarming figures on how much money is being handed out to renewables is courtesy of the World Nuclear Association, "Energy Subsidies and External Costs—Subsidizing Energy," www.world-nuclear.org/info/inf68.html.

p. 204 For all those insights into how the UK's Department for Energy has been busy raising bills, see the Policy Exchange's policy institute publication by Simon Less, "Green Bills: An Analysis of the Projected Levy in Energy Bills," August 17, 2010, http://issuu.com/policyexchange/docs/greenbills.

p. 204 Who emits most carbon dioxide anyway? See http://www.carbonmarketdata.com/cmd/publications/EU%20ETS%202009%20Company%20Rankings%20-%20June%202010.pdf.

p. 206 The ExternE (ExternE Externalities of Energy) study has its own website, www.externe.info/ with links to all its various publications.

p. 207 One key earlier study that ExternE drew on is "Allocation of Environmental Damage from Combined Heat and Power Production between Electricity and Heat—Proposal for an EXTERNE Guideline," by W. Krewitt, A. Lorenzoni, and P. Pirilä, May 1996.

p. 208 The US National Research Council has published a study on the hidden health and environmental costs of energy production and consumption in the USA. It put these costs at $120 billion in 2005. "Hidden Costs of Energy: Unpriced Consequences of Energy Production and Use," http://www.nap.edu/catalog.php?record_id=12794.

*Other data sources and suggestions for further reading for this chapter are:*

Roger H. Bezdek and Robert M. Wendling, "The US Energy Subsidy Scorecard," Issues Online, 2006, www.issues.org/22.

"Chinese-US Deal Opens Opportunities for Nuclear Cooperation," *World Politics Watch,* January 4, 2007.

"Construction of Olkiluoto-3 behind Schedule," *Nucleonics Week,* February 2, 2006.

"Nuclear Power Generation Cost Benefit Analysis," Department of Trade and Industry, London, 2006, www.dti.gov.uk/files/file31938.pdf.

Doerte Fouquet, ed., "Prices for Renewable Energies in Europe: Feed-in Tariffs versus Quota Systems—a Comparison," European Renewable Energies Federation 2006-07, www.eref-europe.org/dls/pdf/2007/eref_price_report_06_07.pdf.

"EDF to Build Flamanville-3, Says First EPR Competitive with CCGT," *Nucleonics Week,* May 11, 2006.

M. Gielecki and J. G. Hewlett, "Commercial Nuclear Electric Power in the United States: Problems and Prospects," *U.S. Energy Information Administration Monthly Energy Review,* August 1994.

Paul Gipe, "Renewable Energy Tariffs in Europe and Elsewhere," 2007.

Management Information Services, "Analysis of Federal Expenditures for Energy Development," 2008.

"Host of Problems Caused Delays at Olkiluoto-3, Regulators Say," *Nucleonics Week,* July 13, 2006.

International Atomic Energy Agency, Power Reactors Information System database, www.iaea.org/programmes/a2/index.html.

International Energy Agency, "Renewables for Power Generation: Status and Prospects," Paris: Organization for Economic Cooperation and Development, 2003.

Nuclear Energy Agency/International Energy Agency, "Projected Costs of Generating Electricity: Update," 2005.

Nuclear Energy Agency, "Nuclear Energy and the Kyoto Protocol," 2002.

OXERA, "Financing the Nuclear Option: Modelling the Costs of New Build," June 2005, www.oxera.com/cmsDocuments/Agenda_June%2005/Financing%20the%20nuclear%20option.pdf.

UBS Investment Research, "More a Question of Politics than Economics," Series: The Future of Nuclear, March 2005.

United States Nuclear Regulatory Commission, "Design Certification Application Review—AP1000," www.nrc.gov/reactors/new-licensing/designcert/ap1000.html 2003.

## AFTERWORD: WHERE NOW?

p. 216 That contrarian assessment of energy costs was published in August 2011 by Dr. Gordon Hughes, "The Myth of Green Jobs," Global Warming Policy Foundation, GWPF Report 3, 2011, http://thegwpf.org/index.php?option=com_acymailing&ctrl=url&urlid=1402&mailid=359&subid=7995.

p. 218 For more on the competing merits of various global transport strategies, see Rail vs. Truck Energy Efficiency, which not only discuss whether "Rail Freight is Usually More Energy-Efficient than Truck?" but also the merits of water-based freight, all at www.lafn.org/~dave/trans/ . . . /rail_vs_truckEE.html.

p. 219 The figures on transport efficiency come from various studies including the Transportation energy efficiency—InterAcademy Council, www.interacademycouncil.net/CMS/ . . . /11924.aspx and "Transport and Carbon Emissions in the United States: The Long View," at www.stanford.academia.edu/ . . . /Transport_and_Carbon_Emissions_in_the_United_ States_The_Long_View.

p. 220 "Japanese Crisis Agenda: The Energy-Saving Measures," *Wall Street Journal* and Bloomberg, August and September 2011, traditionally the months of highest electricity usage in Japan. See Associated Press reports, September 2, 2011.

p. 220 Finally, some interesting tips: "How to Save Money on Gas," www.opentravelinfo.com/ . . . /how-to-save-money-on-gas-29-tips.html.

# INDEX

*Page numbers in bold indicate a glossary entry, items followed by the word "box" indicate the text appears in a text box*